AF835078

PHILOPONUS

On Aristotle

Physics 1.1-3

PHILOPONUS

On Aristotle Physics 1.1-3

Translated by
Catherine Osborne

B L O O M S B U R Y
LONDON • NEW DELHI • NEW YORK • SYDNEY

Bloomsbury Academic
An imprint of Bloomsbury Publishing Plc

50 Bedford Square
London
WC1B 3DP
UK

1385 Broadway
New York
NY 10018
USA

www.bloomsbury.com

Bloomsbury is a registered trade mark of Bloomsbury Publishing Plc

First published in 2006 by Gerald Duckworth & Co. Ltd.
Paperback edition first published 2014

British Library Cataloguing-in-Publication Data
A catalogue record for this book is available from the British Library.

ISBN HB: 978-0-7156-3409-7
PB: 978-1-4725-5769-8
ePDF: 978-1-4725-0131-8

Library of Congress Cataloging-in-Publication Data
A catalog record for this book is available from the Library of Congress.

Acknowledgements
The present translations have been made possible by generous and imaginative funding from the following sources: the National Endowment for the Humanities, Division of Research Programs, an independent federal agency of the USA; the Leverhulme Trust; the British Academy; the Jowett Copyright Trustees; the Royal Society (UK); Centro Internazionale A. Beltrame di Storia dello Spazio e del Tempo (Padua); Mario Mignucci; Liverpool University; the Leventis Foundation; the Arts and Humanities Research Board of the British Academy; the Esmée Fairbairn Charitable Trust; the Henry Brown Trust; Mr and Mrs N. Egon; the Netherlands Organisation for Scientific Research (NWO/GW), Dr Victoria Solomonides, the Cultural Attaché of the Greek Embassy in London. The editor wishes to thank Peter Lautner, Ian Mueller, Edward Hussey, and Mossman Roueché for their comments, John Sellars for preparing the volume for press, and Deborah Blake at Duckworth, who has been the publisher responsible for every volume since the first.

Typeset by Ray Davies

Contents

Abbreviations

CAG = *Commentaria in Aristotelem Graeca*, ed. H. Diels, 23 vols (Berlin: Reimer, 1882-1909)

Charlton = William Charlton, *Aristotle's Physics Books I and II*, Clarendon Aristotle Series (Oxford: Clarendon Press, 1970)

DK = Hermann Diels and Walther Kranz, *Die Fragmente der Vorsokratiker*, 3 vols (Berlin: Weidmann, 1951)

FHSG = W.W. Fortenbaugh, P.M. Huby, R.W. Sharples, and D. Gutas, *Theophrastus of Eresus: Sources for his Life, Writings, Thought, and Influence*, 2 vols (Leiden: Brill, 1992)

Guthrie = W.K.C. Guthrie, *A History of Greek Philosophy*, 6 vols (Cambridge: Cambridge University Press, 1962-81)

KRS = G.S. Kirk, J.E. Raven and M. Schofield, *The Presocratic Philosophers*, 2nd edn (Cambridge: Cambridge University Press, 1983)

L&S = A.A. Long and D.N. Sedley, *The Hellenistic Philosophers*, 2 vols (Cambridge: Cambridge University Press, 1987)

Ross = W.D. Ross, *Aristotle's Physics: A Revised Text with Introduction and Commentary* (Oxford: Clarendon Press, 1936)

Introduction

Aristotle's *Physics*

The work known to us, and to the ancient commentators, as Aristotle's *Physics* is a treatise in eight books on a range of topics to do with the structure of the physical world, and the explanation of motion. The topics are not confined to what we would call physics, if by that we mean the scientific analysis of forces and the explanation of why a body moves. Rather, much of what Aristotle has to say falls under what we would describe as metaphysics: questions about place, about time and the structure of matter, and discussion of how change is possible.

Nevertheless, Aristotle draws a distinction between enquiries that belong to natural philosophy (or *physics*) and those that belong to a meta-level of investigation, which John Philoponus calls 'first philosophy'[1] and we might be inclined to call 'metaphysics'. 'First philosophy' is how I have translated the expression in this volume, with 'metaphysician' for the 'first philosopher', the expert in this branch of study.[2] The fit with our own terminology is not at all exact, however. As we have observed, the philosophical analysis of the structure of time, place, space, and change – as opposed to the empirical investigation of their application – might well belong in metaphysics for us, but Aristotle includes them in natural philosophy, because they are about characteristics of natural things ('adjuncts that belong to all natural things in common' as Philoponus puts it).[3] By contrast, the topics that Aristotle brackets off, to be assigned to first philosophy, are questions about the first principles of a particular discipline such as natural philosophy (or geometry, mathematics, or whatever): that is, the axioms that the subject must take for granted in its investigations. Natural philosophy must begin from some starting points, and those starting points cannot themselves be investigated from within the subject, but must be referred to a different discipline, a meta-discipline that investigates whether the starting points were correct. This meta-discipline, Philoponus tells us, is 'first philosophy'.

This description of an architectonic metaphysics, charged with questioning the axioms of the physicist, does not exactly correspond with Philoponus' initial analysis of the three-fold divisions of theoretical philosophy – into natural philosophy, theological philosophy, and mathematical philosophy.[4] Implicitly, one might suppose, that division assumes that metaphysics is concerned with theological subjects; that is, it assumes that its subject matter is different from the subject matter of physics. Just as mathematical philosophy deals with mathematical objects, so theological investigations treat of divine and eternal objects. And indeed that appears to be how Philoponus understands Aristotle's treatise *Metaphysics*, which, he observes, does its theology in a markedly physical way, basing its theological deductions on physical things (by contrast with Plato whose physics is done theologically in the *Timaeus*).[5] So Philoponus certainly finds theology in the *Metaphysics*. His threefold analysis of the branches of theoretical philosophy does not leave room for a further super-science of first principles, whose task would be to question the principles of mathematics, of natural philosophy, and perhaps of theology as well. It is significant that while Philoponus notes the places where Aristotle refers to such a meta-inquiry,[6] he does not refer to any text where he considers that Aristotle engages in this higher science of first philosophy. He does distinguish it from Dialectic, which he thinks might be what Aristotle means when he refers to 'the common science of all' at *Physics* 185a23.[7]

Yet there clearly is at least one text in which Aristotle engages in that meta-level of discourse, namely the one that is the subject of Philoponus' commentary here. Aristotle's point, when he observes that such questioning of the first principles of natural philosophy does not itself figure as a task for the natural philosopher, is that by rights it should not be a concern in his present treatise. He should, when doing natural philosophy, simply assume the truth of the natural philosophers' basic axioms and derive his conclusions on that basis. He should not begin on the task of demonstrating the truth of those first principles, such as the existence of plurality and motion. Yet he allows himself to indulge in that meta-task, by allowing himself to respond to the Eleatic philosophers, Parmenides, Zeno, and Melissus, who were not strictly speaking natural philosophers, by Aristotle's lights, precisely because they questioned those first principles of physics. Debating with such folk is a task for the first unhypothetical philosophy, not for physics.

So in the text on which Philoponus is commenting, although it comes at the start of a work called *Physics*, Aristotle is not engaged in what he considers to be physics proper. And although Aristotle is engaged in a kind of exploratory debate with earlier thinkers, it is not quite as we would expect. He is not setting out the *endoxa*, with a view to synthesis and selective approval or revision. He is facing a challenge to the very idea that one might engage in physics at all. His task is to establish the first principles of the discipline by showing that Parmenides and Melissus were seriously confused about the notion of being and about the concept of unity. Aristotle's discussion, and Philoponus' extensive analysis and exegesis of it, operate at a level of abstraction that is mind-boggling in the Greek – and even more mind-boggling when rendered into what purports to be English. But our sense that we are dealing with a topic that Aristotle would recognise as part of 'metaphysics' is confirmed by the persistent appeal to themes familiar from the *Metaphysics* and the *Categories*, including the multivocity of the terms 'being' and 'one', the idea that any candidate for being must fall into one of the categories of being, and that only substance can stand on its own without another substantial being to belong to.

Chapters 1 to 3: embarking on the study of nature

Aristotle's attack on Parmenides and Melissus, which is the subject of Chapters 2 and 3 of *Physics* 1, is preceded in the first chapter by some more general considerations about the methodology of enquiry in subjects such as physics. Philoponus' interest is engaged partly by the implicit argument in Aristotle's opening claim to the effect that one should start an enquiry by discovering the subject's principles and causes and elements. Establishing how Aristotle's incomplete argument for this claim could be rendered as a valid syllogism, by identifying its unstated premisses in such a way as to render it conclusive, was evidently a task already prominent in the tradition. Philoponus is responding to an analysis of the reasoning that was to be found in Theophrastus.[8] Here we get a sense that Philoponus, with his somewhat long-winded explorations of how exactly one might construct a syllogistic argument from Aristotle's opening sentence, is part of a tradition of exegesis that goes right back to the generation after Aristotle.

The second problem that engages Philoponus' extended attention in relation to Chapter 1 is a puzzle that still worries scholars today. This is the contradiction, or apparent contradiction, between what Aristotle says here in the *Physics* and what he says in other works (including the *Nicomachean Ethics, Posterior Analytics*, and *Topics*) as to whether the universal is the first thing we grasp or comes later in the process of acquiring understanding. The problem is embedded in a relatively familiar thought, namely that we have to start our enquiries – in physics as in other areas – from what is more familiar to us, even though in the true order of things it may not be more 'knowable' or prior. So the order of investigation does not reflect the logical priority or epistemological value of the knowledge. We start from what is relatively inferior knowledge, but easier to get hold of, and we proceed to a better grade of knowledge, knowledge of things that are genuinely prior and more knowable. This point, made at *Physics* 184a16, is classic Aristotelian methodological dogma.

The problem in this case arises over the status of knowledge of universals. Philoponus explains the issue clearly at 10,23. Normally we should expect Aristotle to say that we start from encounter with the particular, and derive knowledge of the universal at a later stage in the enquiry, while emphasising that the knowledge of the universal is prior and superior in nature. Relative to us the universal is more obscure and comes later in the process of discovery, but in reality the universal is what science is about. Here in the opening chapter of the *Physics*, by contrast, Aristotle surprises us, and Philoponus, with the claim that what is universal is prior and more clear to us, but less clear and posterior in nature. How can we resolve this apparent conflict, so as to show that Aristotle is not contradicting himself but saying something intelligent about the procedure for investigating the principles of nature? That is Philoponus' problem. The answer to it occupies nearly ten pages of the *CAG* edition, followed by a further two pages of close exegesis of the relevant sentences in Aristotle's work.[9]

Philoponus canvasses two possible solutions to rescue Aristotle's consistency, one of which is his own, and his preferred resolution.[10] The alternative, attributed to 'some people' at 11,24, he rejects on the grounds that it fails to fit with Aristotelian theory or the real nature of things. In fact what he shows is that it turns out to amount to something very close to his own preferred solution, though the authors of the supposedly alternative theory do not see it that way since they have missed the point that is crucial to Philoponus' own theory. This crucial point is the distinction between the individual and the indeterminate particular. The individual, Philoponus maintains, is one definite item, such as Socrates; the indeterminate particular is something like 'a human being' or 'an animal', where a number of different individuals can be so described. It is indeterminate and indiscriminate in that it can pick out any one of a great many things, indiscriminately identified under the same description. Philoponus suggests that it is this universality of the indeterminate particular that meets the requisite criteria for being what we encounter first, but being less clear and less knowable in nature. He also makes a convincing case for the idea that when one perceives something as an indeterminate particular in this way, one need not yet have a clear grasp of the genuine universal, clearly articulated as such. That may be something still to be achieved. 'Universal' (*katholou*) in this text has therefore to be understood in a rather specialised way, as referring to the indiscriminate particular, by contrast with the individual.

Philoponus' solution to the apparent contradiction has some merits, although the delivery of it is marred by occasional inconsistency in the use of its own technical terminology. Philoponus himself seems to lose track of which is which between the individual and the particular on page 14, or at least he appears to become careless in his choice of terms, in a section where consistency was crucial. But still, overlooking that brief lapse in his otherwise painstaking treatment of the problem, we may see that his introduction of the notion of an indefinite particular is helpful in capturing what Aristotle ought to be saying, if he is to be saying something both plausibly true and consistent with his claims elsewhere. It is a fine example of the principle of charity applied by a commentator in discussing a difficult text, and bearing fruit for our understanding both of Aristotle's meaning, and of the subject under discussion, which Aristotle had tried to explain in a less than transparent manner.

In Chapter 2 Aristotle turned to a review of his predecessors in the field of natural philosophy. Following his opening remarks about the importance of discovering the principles of the science, he attempted to set out the types of theory found among earlier thinkers as a formal classification according to the number of first principles that a thinker posited to explain the world. If you are doing natural philosophy, Aristotle tells us, you must have either one principle or several, and they must be either motionless or subject to change, and if there is more than one, the number must be finite or infinite. This approach to thinkers of the past is synoptic rather than chronological, and it generates the classic division into monists and pluralists that has stayed with the history of philosophy ever since. It is still the conventional way to classify the Presocratic philosophers, even though the predominant mode of studying these thinkers is now historical and chronological, rather than attempting, as Aristotle was, to chart the range of possible positions that one might hold on a subject.

Philoponus does not question the usefulness of this kind of classificatory review, but rather tries to show that the classification is indeed thorough: that it is 'scientific'[11] and that it is 'lucid' or obvious.[12] That is, Philoponus wants to show that the division is systematic and exhaustive, so that all the options are effectively covered. The alternatives that it canvasses are exhaustive, because they are derived by positing one term and then the negation of that term. Philoponus refers to this kind of exhaustive division as a classification based on contradiction (*diaeresis kat' antiphasin*). The lucidity is a result of the familiarity of the concepts, such as 'one' and 'several', in terms of which the classification is constructed.

In the course of explaining Aristotle's summary, at *Physics* 184b15-25, of the possible positions on how many principles there might be, and of what kind, Philoponus fleshes it out with some names and details. Aristotle does not name the individual philosophers for every arm of the classification because he is going to deal with them in more detail later in the book, from Chapter 4 onwards. Philoponus takes it upon himself to identify one or more thinkers for each of the possible combinations that are realistic, and to give a brief outline of how they are to be understood as falling into that section of the division. In some cases Philoponus passes judgement on the quality of the suggestions that the Presocratic philosopher had offered. In the case of Empedocles, who is both a monist and a pluralist depending which regime is governing the components of the world, he presents a nicely nuanced account which interprets the two worlds, the one and the many, under love and strife, as the intelligible world and the sensible world respectively.[13] He argues that what Empedocles describes as a change of worlds is a description of the soul's move from one world to the other. This brief discussion of Empedocles is a good example of a Neoplatonic interpretation of Empedocles, and is of a piece with Philoponus' Neoplatonic interpretation of the two parts of Parmenides' poem on the same lines, which is offered at 22,4-9 and again at 55,27-57,12 and 65,3-15.

Aristotle is immediately distracted from his task of discussing earlier natural philosophy by a digression to deal with Parmenides and Melissus, who are not to count as *physikoi* due to the fact that they deny the first principles of physics. The remainder of Chapters 2 and 3 of Aristotle's first book are devoted to the refutation of Eleatic monism. Philoponus pauses briefly to explain Aristotle's claim that the dispute with Parmenides and Melissus is not a proper task for physics, and to elucidate the examples of the quadrature of the circle by Hippasus and Antiphon that occur in Aristotle's text,[14] before proceeding to the details of Aristotle's altercation with the monists.

This topic, Aristotle's altercation with the monists, occupies the remainder of the present volume. Philoponus divides it into two main parts: first a section in which Aristotle is considering the question what it would mean to say that being is one, taking the question in its own right and not as a question of exegesis of the ancients' texts; and secondly a section in which Aristotle examines the actual theories or arguments put forward by Parmenides and Melissus. The first of these treats the second half of Chapter 2 (185a20-186a3) and the second covers the whole of Chapter 3, which Philoponus divides into four chunks covering first the discussion of Melissus (186a4-22), second, the discussion of Parmenides (two chunks, 186a22-186b14, 186b14-35), and finally the last part of the chapter, which Philoponus identifies as referring to both Parmenides and Zeno (187a1-10). Much of the discussion of the passage relating to Melissus is an examination of the diagnosis of fallacy in Melissus' argument, which Aristotle claims is eristic. This involves reformulating Melissus' argument into syllogistic form and then showing how it depends upon conversion from the antecedent in a case where the terms are not coextensive, or on equivocation on the term *arkhê*.[15]

The main themes running through Aristotle's attack on the Eleatics are those of the different ways in which one might call something 'one' and the different categories or senses of 'being', according to the classic Aristotelian distinction between substance (*ousia*) and attribute (*sumbebêkos*), and between quantity, quality, etc. within the class of attributes. Philoponus had provided a systematic analysis of the ways in which something might be 'one' in his discussion of *Physics* 185a20. This is put repeatedly to use in explaining the complex explorations of what would happen if one were to suppose that the Eleatics meant that their 'being' was one (really one, or nominally one? generically one or as one individual? and so on), and if one were to suppose that they meant it was one substance or one quality or one quantity and so on.

Philoponus' chief task here is to unpack what is a very compressed and allusive discussion in Aristotle. It is clear at a number of points that even the text is in some dispute, and Philoponus discusses more than one alternative way of construing or punctuating the lemma, seeking the one that makes best sense of the argument.[16]

Philoponus as exegete

As we have noticed, Philoponus operates something recognisable as the principle of charity. He seeks to make good sense of Aristotle's text. He tries to show that Aristotle fulfils his promises and that his project is systematic and appropriate. He explains how the apparent difficulties with reconciling this text with others, or with plausibly true Aristotelian positions, can be eliminated once one adopts a more refined understanding of what is going on in the texts under discussion. He looks for the reading that brings order and intelligibility to Aristotle's text.

Philoponus' own method of procedure is also a model of systematic organisation. He divides the text he is discussing into substantial sections at the points at which Aristotle changes topic. In many cases these major divisions correspond with the chapter divisions or paragraph divisions that are inserted into modern texts, although there is no reason to suppose that Philoponus was reading from a text in which these divisions were already marked out. For each of these major sections he begins with a long expository discussion, explaining what has gone before and why Aristotle is moving in the direction he is now going. This expository discussion, or *protheôria*,[17] is headed by a partial lemma which gives the opening words of the new section – but the discussion which follows is not designed to be exegesis of those words but of the section which those words introduce. At these points I have inserted a heading with the words 'Exposition and Discussion' to show that we are embarking upon one of these introductory discursive passages.

After completing the expository discussion of the whole passage, Philoponus changes gear. He now provides a detailed commentary on issues that arise in the interpretation of particular sentences or phrases in the text of the passage just discussed. This involves going back to the beginning, and heading each new, often very brief, textual comment with the relevant lemma from Aristotle. Usually the first of these mini-lemmata repeats the words that were cited as the main lemma at the start of the expository discussion: the same words are cited as a new lemma, on which to hang detailed treatment of difficulties with its sense or exploration of the implications.[18]

This move from expository discussion of a whole section to detailed exegesis of small lemmata is often not adequately marked in Vitelli's edition in the *CAG*. Because it cites again a lemma from Aristotle that had already headed the long *protheôria* passage immediately preceding this new beginning, Vitelli treats the new lemma, at the start of the *lexis* section, as a continuation of the same passage of Philoponus' work. He marks a new lemma only at the *second* lemma of the new section. Once the divisions are more clearly marked at the *first* lemma of each of these new exegetical sections, Philoponus' lucid and reader-friendly method becomes instantly transparent. For this reason I have entered a second heading marked 'textual analysis and exegesis' at this point and numbered the headings to show the dependence of these sections under the main expository sections that precede them.

These detailed exegetical sections support and defend the points made in the preceding expository section, so that one will often get two treatments of a topic separated by some pages. The topic appears first in the general expository discussion, which explains why it matters and what is going on in the passage as a whole, and then it appears again in the detailed exegesis, when Philoponus shows how he can get that meaning from particular sentences in the text and why the wording matters. For an example of this double treatment of a topic, see for instance the comments on why debate with the Eleatics has some attractions because it has something philosophical about it, which is briefly mentioned in the expository discussion at 28,25, and then treated again in the exegetical analysis of the lemma in question, at 32,15.

Originality and dependence

The principle of charity is evident in the fact that Philoponus tries to make good sense of Aristotle's project and to show it in its best light. However, he continues to maintain a certain critical distance: he is not merely telling his readers what Aristotle was saying, nor does he try to make Aristotle into John Philoponus or into a paid up Aristotelian Neoplatonist.

Doubtless the critical distance is encouraged by the presence of a tradition of other commentators and exegetes whose work Philoponus knows and with whom he needs on occasion to disagree. However Philoponus wears his scholarship lightly: by contrast with Simplicius, he very rarely refers to the previous interpretations and even more rarely does he give the name of the author of some rival interpretation that he is discussing. Just occasionally he protests that something 'some people' say must be wrong, as for instance when he rejects the suggestion – by someone Philoponus does not name (probably Alexander of Aphrodisias) – that Aristotle's comment at *Physics* 185b28 was an attack on Plato.[19] Philoponus tends to proffer his own view, where it is original to him, as a kind of afterthought, following a discussion of rival views in the literature. He concludes by saying 'but we say …'[20] or 'but I say' and then offering a position that appears to be a new contribution.[21] Philoponus offers his own position about a range of matters, including disputes about how best to formalise Aristotle's informal arguments, attempts to identify whom Aristotle is attacking at a particular point, or rival diagnoses of the fallacies in the Eleatic arguments.

But besides disagreements with previous exegetes, over the meaning of Aristotle's text or the identity of the characters under discussion in Aristotle's allusive text, Philoponus also keeps a philosophical distance between himself and Aristotle. Philoponus does not think that Aristotle is infallible, and at various points he offers objections to Aristotle's own theoretical positions, or alternative solutions to the problems that Aristotle raised. An example of the latter kind of response to Aristotle occurs at the very end of the portion of text included in this book, when Philoponus adds a suggestion of his own about the simplest way to refute Parmenides without the need to posit the existence of *not-being*.[22] The implication here is that we do not need to adopt anything so complicated as Aristotle's treatment at *Physics* 187a3-9, invoking the idea of 'not-being-something', to show that there is a kind of not-being that Parmenides' argument does not eliminate. Philoponus suggests that his own solution reaches the required result equally effectively.

Philoponus' objections to significant parts of Aristotelian theory have been more widely discussed, and have become an issue in relation to the dating of the commentary on the *Physics*. The two issues on which Philoponus tends to disagree with Aristotle are (a) the essential nature of body and (b) the origin of the world and its eternity. I do not find myself convinced by the conventional view that the *Physics* commentary displays an uncritical stage in Philoponus' thought, before he had begun to dissent from the Aristotelian positions in these areas.

On the nature of body there are two key passages in the text translated here. The first is at 4,26-35. Here Philoponus is trying to show that all physical things are composite, and he urges that this must be true even of things which are by definition simple if they are the kind of thing that can only occur as attributes of bodies. Philoponus observes that Aristotle has an antipathy to the separate existence of forms, so that if an attribute can only occur in a body, there can be no simple form of it in existence, except in the mind. So any reality that it has will be as part of a composite in which it is present in a body. The rejection of Platonic Forms makes it impossible for attributes to have reality by themselves.

Philoponus himself may not be so negative about Platonic Forms, but he does not voice his dissent explicitly here. Instead he observes that exactly the same situation applies in the case of body itself as applies in the case of the other forms: the form of body is also a simple attribute, he says, namely three-dimensionality. Its own proper definition is just that. Matter is not part of what it is to be body. However, in order for body to occur in nature it must be in a substrate, namely matter. This means that any body that has actual physical existence will be a composite of form and matter. But this is not something peculiar to body. It remains true that *anything* will be composite when it is considered as something real and actually existent, even if its form is simple.

Why does Philoponus mention the form of body here? It seems that the original point (which may be inspired by something in Theophrastus)[23] was that all physical capacities and faculties exist in a substrate and are therefore composite when they actually occur in reality. Philoponus digresses to show that the same is true of body. It seems that he is conscious of the fact that he wants to say that body in itself is simple, as regards the definition of its essence, and he wants to show how one can still hold that all bodies that you actually meet in real life are composites and include matter. This is no more essential to them than is the matter of an eye essential to what sight is. So when body is considered as a form or according to its definition it is simple: it is just three-dimensionality. This remains true even though all real bodies occur in matter.

Philoponus returns to the identity of three-dimensional extension and body at 38,19. Philoponus suggests that we ought to investigate why Aristotle says that infinity occurs as an attribute of substance only *per accidens* in so far as it partakes of quantity, and that substance cannot be infinite *per se*. The focus is on the accidental relation between substance and quantity. Aristotle is assuming that quantity is an accidental attribute of substance. Philoponus finds this worthy of investigation, because if you hold, as Philoponus does, that body is defined as three-dimensional extension, then it seems to be a magnitude, and hence to be quantity in its very being. And if body is a substance this should mean that it is perfectly possible for a substance to be quantity not just *per accidens* but in itself, and hence it could, in principle, be infinite in quantity without that being an additional attribute that it has *per accidens*. Or so it seems.

Philoponus enlarges upon his notion that body is essentially a magnitude by drawing a distinction between essential and non-essential properties of a thing. He suggests that one can have an essential magnitude in the same way as one has an essential property, so that the essential magnitude would have a relation with the matter that is comparable with the relation that the essential qualities of the elements have with their matter (presumably the connection between fire and hot, and so on).[24] So similarly there will be three-dimensional extension in any body, as a *per se* property of it.

This is the thought that requires investigation. There are however two responses to it which show that it is not a fruitful thought after all. The first is that Aristotle does not hold this view of body, and hence it is not surprising that he does not regard magnitude as belonging to the body *per se*. This solves the problem of why Aristotle says that infinity can only belong to something in so far as it partakes in quantity. Aristotle, unlike Philoponus, does not think that body is simply quantity. 'If body is not a simple thing for Aristotle', Philoponus says, 'but is itself also composed of prime matter and quantity (i.e. what is extended in three dimensions), it is clear that body is not simply quantity'.

In this way we can understand what Aristotle is up to even if we do not agree with his analysis of body. But Philoponus has more to say, because even on his own view it will turn out that infinity cannot be a *per se* property of body.[25] Even if we do hold that three-dimensionality is the defining differentia of body, and hence essentially what it is, it still will not follow that any specific size is essential to what it is to be body. The actual size of a body is an accidental property and is not constitutive of the essence of body as such (39,20). So once we distinguish three-dimensional extension from the particular extent or size of something, we see that 'infinite' is a particular size like 'three cubits long'. And this does not belong *per se* to body as such. So it transpires that we must agree with Aristotle that infinity does not occur as a *per se* attribute of substance, even if we hold, with Philoponus, that the essence of body is the three-dimensional.

It seems that Philoponus is committed in both these passages to a consistent view which he takes to be distinct from Aristotle's own view. He holds that body is to be defined simply as three-dimensional extension, and that matter is not part of what it is to be body. However he holds that the form can only exist in reality when it is instantiated in matter, so for every body that actually exists there will need to be a substrate to which body belongs as form. Bodies, then, are composites of matter and form, but body is just the form and is simple. But while body is three-dimensionality, only particular bodies have size, so that any particular size is an attribute of a composite body, not body as such. At 39,3 Philoponus suggests that a whole dedicated treatise would be needed on the question whether three-dimensionality is an essential property of body. It is unclear whether he is alluding to the (planned?) *De Aeternitate Mundi contra Proclum* in which this doctrine is repeated several times,[26] or simply indicating that it is a topic that he would like to write about sometime. So these passages of *in Phys.* appear to hold the developed doctrine that extension is the essential differentia of body, but not the denial of prime matter as a necessary first substrate for bodies, which scholars have identified as new in the *De Aeternitate*.

On the origin of the world and its eternity there are relevant passages at 1,23-5; 9,29; 15,30; 16,1-10; 16,25-30; and 54,10-55,26. The crucial thing, in deciding whether Philoponus already holds anti-Aristotelian views, is to distinguish whether he is merely offering exposition and summary of the assumptions in Aristotle's text, in order to explain why Aristotle says what he says, or whether he is standing at a critical distance and offering reflections of his own on these topics. In the first case there is no necessity for Philoponus to mark his disagreement, even if he does disagree, where the sole purpose of his discussion is to explain Aristotle's take on things.

The first passage in our list above (1,23-5) seems to be such a candidate, perhaps intended as mere summary of Aristotelian teaching. In his opening pages Philoponus is listing the different works in which Aristotle deals with different parts of natural philosophy, and he blithely lists the *De Caelo* as a work concerned with the properties or adjuncts that are peculiar to eternal things in particular. The fact that at the time of writing Philoponus might not hold that the items in question are eternal in the way that Aristotle supposed need not prevent him from writing in this way, just as a modern scholar might describe the content of the *De Caelo* in the same way, without endorsing the views expressed there at all.

The second passage (9,29) similarly purports to show what Aristotle himself should have done, given his belief in a fifth element. The complaint is an *ad hominem* one. If one were Aristotle, and one were trying to produce a scientific proof about the phases of the moon, then the didactic procedure would dictate that one would first show that it was made of the fifth element and then show how that would entail that it was spherical and would have phases as we find them. The fact (if it were a fact) that Philoponus himself did not accept the doctrine of the fifth element would be irrelevant to his example of what a correct Aristotelian scientific proof on the subject would look like. So we cannot read off from this passage any conclusion about where Philoponus stood on the subject of the fifth element. But it is worth noting that Philoponus is quite clear that it is actually impossible to establish (from down here) what stuff the moon is made of.[27] Perhaps that was precisely his point.

The next two passages (15,30 and 16,1-10) are continuous. Philoponus is illustrating the idea of dealing with a topic that is inarticulate and indiscriminate, and he suggests that talking about 'matter' is such a topic, because there are differences between the matter of different kinds of things: the matter of plants is different from the matter of animals; the matter of things on the ground is different from the matter of atmospheric phenomena. And the matter of eternal heavenly things is different from the matter of things that come to be. The last of these is the first in Philoponus' list, and we immediately think of Aristotle's fifth element. But the other examples show us that we do not need to think of any hugely technical distinction between these kinds of matter. And even if it is an allusion to the idea that there is a fifth element, it still remains true that Philoponus is using an example of how *Aristotle* would agree that the notion of matter is used indiscriminately if it is used to cover all these different kinds of matter. Once again it does not carry any implication that Philoponus would endorse exactly this account of what heavenly bodies are like, or what they are made of.

On the next page (16,1-10), however, Philoponus asserts, apparently in his own voice, that there is not a unitary nature to matter, and that even the prime matter of heavenly things is distinct from that of things involved in generation. This repays a moment's reflection. If prime matter is a formless stuff, and the elements are already a compound of form and matter, then one would expect the fifth element to be a different form, but why should it be in a different kind of matter? The point is related to one about distinguishing what is the proper and proximate principle of each thing, not some supposed principle at too high a level of generality to be explanatory. Is Philoponus endorsing the idea that the heavenly things are made out of a fifth element? Or is he merely suggesting that for any subject of enquiry one needs to find the peculiar principles of that thing, and that it is not sufficient just to take matter, as though that were an uncontroversial single stuff. The fifth element, whether one accepts it or not, is a useful dialectical tool to show just how useless the notion of 'matter' is if we fail to discriminate between different kinds.

Philoponus continues his attack on the notion that talk of prime matter is intelligible or helpful in the next adjacent passage, 16,25-30. The whole discussion is focused round the issue of what it is for something to be 'more knowable and clearer in nature' (or 'to nature'). Philoponus is asking how nature relates to prime matter: whether prime matter could be one of the things that is clear and knowable in nature, or whether it is one of the things that is confused and indiscriminate. His point is that prime matter is not a clear and knowable first principle in nature. Besides the previous claim that it is too generic and fails to mark important distinctions between different kinds of matter, he now tries to make a point about the order of creation. Assuming that nature knows what she makes, we must ask, 'Does nature make prime matter?'. Philoponus answers 'no'. Nature makes proximate matter, the matter for each of the things that are created, but she does not make prime matter. Prime matter, he says, is supplied from the start.

The passage leaves it unclear which of two things Philoponus means, when he says that prime matter is supplied from the start. Does he mean that it is created, but created first and not by nature, or does he mean that it is without origin and eternal? The former might be compatible with a divine creation *ex nihilo*, and could therefore be made compatible with Philoponus' mature view that the world was created. But in fact we ought to read the passage in the latter sense, as a claim that prime matter is pre-existent and eternal. This will become clearer when we compare the passage with the later one at 54,10-55,26, but first we should note that within this context it makes better sense to take it that way.

In order for the argument to work, we have to reach the conclusion that there is no standard of knowledge such that the prime matter is supremely knowable and more knowable than the derivative products, as it would be if it was what nature primarily made. Philoponus wants to say that prime matter is not more knowable in nature because it is not something that nature, the creator, made. So if there were some divine creator who made first matter first, that would undermine the point, for now the God who made matter would seem to provide the ultimate standard for knowledge. Thus prime matter would seem to be what God would know if God were author of reality as such. By creating matter *ex nihilo* he would make it the first and most knowable thing to him. It would be merely specious to say that since he is not identified with nature, the things thus created are not more knowable to nature. So, in so far as the argument is a good argument to show that prime matter cannot be more knowable in nature, it seems that Philoponus needs to say that there really is no creator of prime matter. In this way I think we can see that the claim that 'prime matter was provided from the start' ought to mean 'prime matter is uncreated and eternal, having no origin at all', if Philoponus is not to undermine the effectiveness of his argument.

The picture of a pre-existent formless uncreated matter, out of which a demiurge creates the elements and the proximate matter for the creation of individual things, is reminiscent of Plato's *Timaeus*. But it seems that it is also a picture that Philoponus is prepared to canvass as one of two possible accounts of the origin of the universe. Philoponus offers a choice between two accounts of the origin of matter at 54,10-55,26, and one of these appears to be identical to the one we have just encountered, namely the idea that matter as such has no origin but is 'supplied in advance'.

Here at 54,10 Philoponus decides to question Melissus' claim that being could not come into being. Aristotle, he observes, did not question that axiom, because he thought the same as Melissus. Since Aristotle too thought that the universe had no beginning, he saw no reason to pick holes in the claim that being could not have a beginning, but Philoponus is not convinced. He is prepared to envisage the possibility that being could have come from not-being, that what is could have started *ex nihilo*. The passage from 54,10 to 55,26 explores how that could be true.

The first suggestion at 54,16 is that there *need not* be a pre-existent material. We can talk of something having the potential to come into being even when there is no antecedent material, providing an efficient cause exists first. So one can have a case of coming to be where the matter is not supplied in advance, although there must be something that exists in advance because there has to be a cause. So this is not coming to be from nothing. It is coming to be where the matter is brought into being at the same time as the product. Indeed Philoponus thinks that we should not use the language of potentiality where there is pre-existent matter, unless the matter is shown to be uncreated in itself. Here it seems that Philoponus leaves space for two alternatives that will allow us to speak of potential being: either if the things that are to come into being are brought into being by an efficient cause who creates matter at the same time, or where there is pre-existent matter that is supplied in advance and is without origin and indestructible in its very specification. These two options are repeated at 54,30, where

Philoponus again urges that if we speak of potentiality we ought to be speaking of originating from not-being, unless we have shown that matter itself is essentially ungenerated.

At 54,31-55,5 Philoponus indicates that he takes himself to have shown that coming to be does not necessarily presuppose a material cause 'from which' the new item comes, but that coming to be 'from non-being' is allowed where the matter is called into being along with the forms. Then, at 55,5-20 he questions the coherence of talking of 'being' *simpliciter* coming into being, given that 'being' is not a single class but distributed across the categories. So there is no one entity that we mean when we speak of 'being'. This makes nonsense of asking, about 'being' in general, whether it came into being from non-being. On the other hand it does make sense to ask of body, or of the world, whether it came into being, and if so, whether from being or from non-being. This question makes sense, and is the question Philoponus intends to be asking. He proposes two answers to it at 55,22: (a) that it (the world or body) came from what is, if matter is supplied in advance, or (b) that it came from non-being if matter was brought into existence at the same time as the form. 'Such are the responses to be advanced on our part, for the present, against the claim that being did not come into being ...'.

These two answers are the two answers that we associate with Philoponus. One is the answer we have found him adumbrating above, at 16,25-30, that there is a pre-existent material cause, prime matter, that is eternal and ungenerated, which can correctly be said to have the potential to become what it later becomes, and from which the things come into being (so that being can be said to come from being). The other is the answer we associate with his later works, one that looks more thoroughly Christianised, that matter itself was created *ex nihilo*, and being came out of non-being when the world was created (though not without a creator as efficient cause).

Why does Philoponus offer both answers here (without adjudicating), and only the first one at the earlier passage? Three stories seem possible. Either Philoponus started his work on the *Physics* with the first thesis in mind – a thesis perhaps absorbed from the Neoplatonism of his mentors – and delivered this as his own opinion at 16,25-30; then when he reaches the passage in which Aristotle fails to consider whether being might come from non-being, Philoponus himself is prompted to see that the dogma that being cannot come from non-being is necessarily true only if it can be shown that matter is unoriginate. Finding himself unable to demonstrate that (which he had hitherto supposed true) he presents two alternatives, leaving open the possibility that it might still be possible to demonstrate the eternity of matter but showing that if that is not possible one must suppose that being could come from non-being. This would be to suppose that Philoponus revises his commitment to the first thesis in the course of composing this work.

A second option is to suppose that there has been a second redaction of the present commentary, and Philoponus has made partial revisions to make allowance for his later thesis. This too envisages a change in Philoponus' view, but a change that took place after the first edition of the commentary had been completed. This is Verrycken's hypothesis.[28] It makes it easier to explain the cross-reference to a work in which Philoponus thinks he has challenged the idea that being did not come into being. On the other hand it is hard to explain what the point of the passage would have been before the second redaction, if it proffered only thesis (a) which would endorse, rather than challenge, the claim that being never comes from non-being.

The third option is to suppose that there is no change of view but that Philoponus continues to think that both options are 'not impossible' as he puts it. Why then would he canvass only the first one at 16,25-30? Perhaps because on neither view does anyone make *prime matter*, so that the relevant point at 16,25-30 is the denial that nature makes prime matter. If this is so, then in so far as we presuppose the existence of prime matter, it will be as something that is not made and does not have an origin. This disposes of the thought that prime matter is what is better known to nature, the relevant point at 16,25-30. On the other hand, if we take the second view and suppose that matter is brought into being at creation, it will surely be as actual matter, the matter of the world, that it emerges; and hence it is not a relevant challenge to the claim that no one makes prime matter.

Why should Philoponus want to leave open the option that prime matter is uncreated and pre-existent, and might potentially be proven to be so, even after he has concluded that creation *ex nihilo* is not ruled out? One possibility is simply that he is unable to prove that it is not so. At 54,29-30 he rather implies that the burden of proof is on someone who defends the eternity of matter. At 55,22-4 the optatives imply that both theses are equally hypothetical and awaiting conclusive proof.

If it is true that both theses are tentative alternatives for which Philoponus has as yet no decisive argument, it is tempting to think that his uncertainty might relate to the question about the nature of prime matter and body. If he is unsure whether matter is a further item besides the three-dimensionality of bodies, and whether three-dimensionality should be said to be created when body is created, or should be considered to be like prime matter, as an antecedent substrate, a given even for the creator, then he might be unsure whether to say that there was something that was in itself uncreated, namely three-dimensionality that is the *sine qua non* of body, or that this too was brought into reality just in so far as body was created that had three dimensions. Such an antecedent substrate, three-dimensional extension, would have some claim to have the potential to become what the creation would make of it.

Philoponus and Presocratic philosophy

Philoponus is interesting in his own right as a critic of Aristotle and of the Aristotelian account of the origins of the world. He is also interesting, as are the other commentators, for the evidence that he passes to us about lost works. Although Simplicius is a richer source of material on the Presocratic philosophers, Philoponus' musings on whether Aristotle has correctly responded to Parmenides, Melissus, and the other Presocratics are not without interest.

At 65,10 Philoponus has been tracking Aristotle's exploration of what Parmenides might mean when he says that being is one. The question is whether he means that being is one substance or one attribute. Philoponus summarises the absurdities that are supposed to follow on the assumption that there is only substance (but that it has attributes which must accordingly be non-being). Philoponus suddenly stands back from Aristotle's text and observes that rather than just follow Aristotle's lead, we ought to 'listen sympathetically to Parmenides' own words, and to grasp his own meaning from his original text'. The thought is a promising one, and Philoponus presents a few lines of text (lines 4 and 25 of fragment 8 and a variant version of what looks like line 5). Yet the lines seem to be insufficient to enable us to feel that we are really in a position to 'grasp his meaning from his original text'. Arguably Philoponus means that he himself has turned to the text and tried to listen sympathetically, even if he does not give us the material from which to do so ourselves. He gives us, rather, the proof texts that he has selected to illustrate what he takes to be a charitable reading of Parmenides: selected lines that show Parmenides speaking as though there is some kind of plurality in being.[29] This is then married with the thought that Parmenides was not talking about physical things anyway (a thought perhaps reinforced by Aristotle's claim in Chapter 2 that the monists were not natural philosophers) and generates the conclusion that Parmenides was paying attention to the unity and immutability of intelligible things and this was why he stressed the idea of unity and unshakability. The suggestion is, effectively, that Aristotle has missed the whole point, and that his extensive analysis of whether Parmenides was talking about one attribute or one substance, was all beside the point. In fact the theory was about the unity of the intelligible world, the world of forms.

Whether this Neoplatonist reading of Parmenides is Philoponus' own original work or one he learnt in the Schools, it is one that he applies consistently. It picks up on the themes canvassed at 22,4-9 and 55,27-57,12. At 22,4-9 Philoponus disputes Aristotle's literal reading of Parmenides and claims that the *Towards Seeming* is an expression of Parmenides' own views, not just a summary of mistaken opinions. Rather it is a correct and realistic account of sensibles or opinables, while *Towards Truth* is about the world of intelligibles. The terminology is explicitly linked to the *Timaeus*. At 55,27-57,12 again Philoponus implies that he is going to do some independent work on the meaning of Parmenides' text, citing Aristotle's claim that they are not natural philosophers and claiming that he has evidence from the text for that being correct – 'and indeed the men's own writings show that'. In fact the treatment is unspecific, and appeals again to the Platonic terminology, including the term 'opinables'. What follows is largely glossed and explained in such a way as to bring it into line with Plato: 'One should be aware that they called everything intelligible "being" and what is perceptible "becoming". Indeed that is Plato's terminology too, inherited from him [sc. Parmenides].'

As a source of good quotations and reliable testimonia such treatment of the Presocratics is merely frustrating. This is not a rich hunting ground for authentic 'fragments'. But as a meditation on the continuity between Eleatic and Platonic themes it is by no means uninspired, and is at least as enlightening as the heavily Aristotelianised Parmenides of the text he is responding to. Perhaps we should see his frustration emerging in that protest at 65,10 that we ought to listen sympathetically to what Parmenides himself said in his text. We must not read it through Aristotle's pedantry, he is saying. We must not fall into that over-literal question about whether it was 'one' as real or generic, whether it was 'being' as substance or accident. We should instead see that Parmenides' theory is a vision of the world of intelligible forms, by contrast with the world of perceptible plurality. Philoponus resists the Aristotelian model that he is condemned to annotate, and allows us occasionally to see another way of reading these texts, one that is arguably more attuned to the poetic imagery of Parmenides' own writings.

Besides these Neoplatonic readings of Parmenides and Empedocles, Philoponus provides brief allusions to a wide range of early natural philosophers, but little detailed information on any of them. Some of his comments seem ill-informed, particularly with regard to Anaximander (23,14-28), where modern scholarship tends to doubt the Aristotelian tradition to which Philoponus is indebted. There is a brief but rather inadequate explanation of Anaxagoras' theory of homoiomeries (24,24-25,4) and a short discussion of how Democritus' atoms differed from one another (25,14) interspersed with further reflections on Anaxagoras (26,9). Philoponus' grasp of the history of Presocratic philosophy is mainly competent enough to permit him to identify possible candidates for the various theories, but is not immensely informative beyond that point.

Notes

1. Or 'first unhypothetical philosophy'; see *in Phys.* 27,7; 32,16.
2. 27,20; 28,5.
3. 1,22-3.
4. 1,6-7.

5. 5,24-5.
6. Particularly *Phys.* 184b25 and 185a14-20.
7. 27,11.
8. 4,8-11.
9. Sections 1.2.1 and 1.2.2 in the translation below.
10. 10,28-11,23.
11. *epistêmonikê*, 20,24.25.
12. *enargês*, 20,24.26.
13. 24,3-23.
14. 26,21-32,3.
15. 50,30-53,15.
16. For different options on punctuating the sentences, see for instance the discussion of *Phys.* 186b4 at 69,15, and the discussion of *Phys.* 186b14 at 75,5. For problems regarding how to divide the words, see the discussion of *Phys.* 186a16 at 61,2.
17. Philoponus uses the term *protheôria* for the expository discussion in his *De Anima* commentary (which also follows this method of double discussion); see *in DA* 424,4.13.
18. This method of dividing the discussion into two approaches, one expository and discursive and the other textual, became standard practice in the later commentators Olympiodorus, Elias, David, and Stephanus. The second, textual, discussion is known as the *lexis* or *exêgêsis tês lexeôs* (see Philoponus *in DA* 121,10; 124,25). See also Evrard, *L'école d'Olympiodore et la composition du 'commentaire à la physique' de Jean Philopone*; Sorabji, 'The Ancient Commentators on Aristotle', 8 and n. 40; and forthcoming volumes of the *De Anima* commentary in this series. I am grateful to Richard Sorabji for alerting me to these earlier and forthcoming discussions of Philoponus' method.
19. 49,19-20.
20. *hêmeis de phamen*, translated here 'on the other hand our view is …'; *legomen oun*, translated 'our first answer is …'; *eipomen de*, translated 'but our view has been …'; or *phêmi de egô*, translated 'in my own opinion'.
21. See for example 10,23-30; 11,24-12,3; 36,4-15; 66,7. The last of these may not be intended as an assertion of originality but rather a reference back to an earlier statement of the view.
22. 86,12.
23. See 4,8 and notes ad loc.
24. On the relation between formless prime matter and the form of body see also 16,1-20.
25. 39,15-29.
26. This work is currently being translated into English in this series; the first two volumes to appear are *Against Proclus on the Eternity of the World 1-5*, trans. M. Share (London & Ithaca, NY: Duckworth & Cornell, 2005) and *Against Proclus on the Eternity of the World 6-8*, trans. M. Share (London & Ithaca, NY: Duckworth & Cornell, 2005).
27. 10,2-3.
28. See Verrycken, 'The Development of Philoponus' Thought and its Chronology'.
29. It is not clear why line 4 is included, unless the reference to 'sole in kind' is meant to indicate that the unity of the one is generic unity rather than substantial unity.

Textual Emendations

The translation in this volume follows the text printed in *CAG* vol. 16, *Ioannis Philoponi in Aristotelis Physicorum Libros Tres Priores Commentaria*, ed. H. Vitelli (Berlin: Reimer, 1887), but note the following deviations:

6,15-16: Reading *mia men oun arkhê autê, tên hulên legôn*, as suggested by Vitelli in his apparatus (ad loc.) from Aristotle

22,17: Omitting the word *apeiron*, entered in square brackets by Vitelli. If *apeiron* were retained the sentence would read: 'but Melissus, who was also talking about the same things, as I said, held that it was one and motionless and unlimited'

28,9: Omitting *anankê pasa* excluded in square brackets by Vitelli on the grounds that it has crept in from the later occurrence in line 11

32,23-5: Following Vitelli's suggestion in the apparatus, I am reading *epeidê* in the lacuna before *ekhei* in line 23 and deleting it from line 25

36,23: I have found it impossible to make sense of the reading *kath' hauto*, and have translated on the basis that the text should read *hôsper hai deka katêgoriai kat' auto to on hen eisin*

37,14-15: Retaining the sentence *epitasis esti tou atopou to adunaton* ('impossible is a stronger form of absurd'), which Vitelli places in square brackets. See note ad loc.

39,16: *hêitini* is read as a single word

39,18: *sunekhê* should be read without iota subscript

41,26: Reading *autai men gar hai ...* (for the ideas themselves) rather than *hautai men gar hai ...* (for these things, the ideas)

47,22: The sense demands the translation I have given. It is obtained by reading (as Vitelli suggests in the apparatus) *apêrtêmenê* (disconnected) in place of *anêrtêmenê* (connected)

80,9: Assuming *athakatos* is a misprint for *athanatos*

Philoponus

On Aristotle

Physics 1.1-3

Translation

Commentary of John Philoponus on Book 1 of Aristotle's *Physics* 1,1

<Introduction: Aristotle and the study of nature>

Even though according to Aristotle there are two parts to philosophy, theoretical and practical, Aristotle himself devoted virtually all his attention to the theoretical part. For it is by theoretical philosophy that we discover the nature of things as things that have being. Theoretical philosophy is divided into three parts: (a) natural philosophy, (b) theological philosophy, (c) mathematical philosophy; of these it was with natural philosophy that Aristotle was most concerned, on the grounds that it was more commensurate with our nature, and that theological and mathematical philosophy had been practised by many people even before Aristotle, while the same was not true of natural philosophy. For this reason Aristotle wrote numerous works of natural philosophy, and distributed his abundant writings over pretty well the whole range of natural topics. In order to illustrate this it would be a good thing if we made a list of the adjuncts that accompany natural things; in this way we shall see that Aristotle's works were also divided in accordance with the things of nature.

Some adjuncts are common to all things; others accompany some in particular. Of the ones that accompany some in particular, some belong to eternal things in particular, others to those involved in generation and corruption. Of those belonging to things involved in generation and corruption, some belong in particular to things above the ground, others to things on the ground; and of those that belong to things on the ground, some belong in particular to lifeless things, others to things that have life; and of those that belong to things that have life, some belong in particular to things that have senses, others to things without senses.

Aristotle, then, wrote about things that belong to all natural things in common, namely in the work before us; about those that belong to eternal things in particular in the *De Caelo;* and about adjuncts that universally accompany all things involved in generation and corruption in the *De Generatione et Corruptione,* whereas in the
Meteorologica and *On Mines* (that is the subject of the fourth book of 2,1
the *Meteorologica*), in which he taught on matters that appertain to lifeless things, he wrote about those <adjuncts> that belong to atmospherical phenomena in particular. As for the corpus of works on animals and plants, some are about adjuncts that accompany them as living but not having perception, i.e. the works on plants, while

others are about those that accompany them as living and perceiving, i.e. all the treatises and investigations on animals. Some adjuncts go with the animals as a whole, some with parts of them. The adjuncts that go with them as a whole are discussed in the work *On Animals*,[1] those that go with their parts in the *De Partibus Animalium* and *De Motu Animalium*. Also *De Somno et Vigilantia*, *De Vita et Morte* and similar works[2] relate to the study of animals, and furthermore the *De Anima*. Such is the sum total of Aristotle's works on nature.[3]

The book we have before us is, as I said, about the adjuncts that accompany all natural things in common, so that it is appropriate that he named the treatise *Phusikê*.[4] There are five of these adjuncts: matter, form, place, time, motion. In the first four books he lectures on the first four subjects, and in the last four books on motion; for the account of motion is complicated and it has many adjuncts. Hence he often calls the whole work 'On motion', saying 'as we said in the books On motion', that is the *Physics*, naming the whole treatise from the part.

According to some of the natural philosophers,[5] the void and infinity are also adjuncts that accompany all natural things. For Democritus[6] used to say that all natural things are made up of atoms and the void, and then again Anaxagoras[7] said that the homoiomeries[8] in each thing were infinite. And indeed Democritus suggested that the atoms were infinite. But Aristotle demonstrates in the fourth book of the present treatise,[9] that it is nowhere possible for void to exist, and he shows that infinity cannot exist in actuality, nor all at the same time and in the same way, but rather only in potentiality and constructed part by part;[10] for there are things in the universe such that they cannot coexist all together but have subsisted part by part. For example we say that the day exists, not in virtue of the whole existing at once but in virtue of its existing part by part. Similarly we say that a contest has taken place not by the whole existing together but part by part, the boxing, say, or the wrestling, or something else. In this way Aristotle shows that infinity is in one sense existent, in another sense not; for he shows that it does not exist as a whole, but part by part in virtue of things continually coming into being to infinity.

Privation is also an adjunct of natural things, but not of all but only of those involved in generation and corruption. For these by nature partake partly of form and partly of privation. So much for the
3,1 adjuncts that accompany natural things.

In the first book Aristotle dismantles the opinions of his predecessors about the principles of nature, and lectures on form and matter, and in the second on form (though he also discusses matter, just as he discussed form as well in the first book; but there is more about matter than form in the first book, just as there is more about form than matter in the second). In the third he lectures on motion and infinity, in the fourth on place, time, and void, and in the remaining four books on motion and all its adjuncts. Indeed Aristotle himself, whenever he wants to refer to these books, says 'we said in the books On motion'.[11]

<1 Methods of enquiry: Book 1 Chapter 1, 184a10-184b14>

<1.1.1 Exposition and discussion: concerning Aristotle's method, 184a10-16>

> **184a10** Since knowledge and understanding come about in all the disciplines that have principles or causes or elements from knowing[12] these <principles or causes or elements>.

It is Aristotle's custom to begin his works from certain common assumptions. Hence in the *Metaphysics*[13] he started from a common agreement: 'All human beings', he says, 'naturally long to know. The evidence is their love of the senses'; and in the *Posterior Analytics*,[14] 'All teaching and learning come from pre-existing knowledge'; and in the *Ethics*,[15] 'Every technical or scientific discipline, and similarly every action and choice, seems to aim at some good'. Here too in the same way it is from a common assumption that he makes his beginning. The assumption is as follows: every science that has principles or causes or elements becomes known when the principles and causes and elements come to be known. Since then, he says, there are principles and causes and elements of physical things, we shall come to know these things when we know their principles and causes and elements. As a whole the reasoning[16] is this: in natural science there are principles and causes and elements; every science in which there are principles and causes and elements becomes known when the principles and causes and elements have become known; therefore natural science also will become known when the principles and causes and elements have become known. Such is the reasoning in its entirety, but for brevity Aristotle sets down only the major premiss (the one that says that every discipline that has principles or causes or elements becomes known when the principles and causes and 4,1 elements have become known) and omits the minor premiss that physical theory has principles and causes and elements. He also omits the conclusion, and sets down only what follows from the conclusion: for the conclusion was 'therefore physical theory also becomes known when its principles are known'. He gives the consequence of this, namely that since physical theory becomes known when its principles are known (which was the conclusion of the syllogism) we should try to define the principles of physical theory.

This is how Aristotle wrote; but Theophrastus[17] set out the whole syllogism in his own treatise *On Nature*, thinking an explanation was also required for the minor premiss (that physical theory has principles and causes and elements); for this premiss is not self-evident. Theophrastus therefore supports it in the following way: physical things, he says, are either bodies or have their existence in bodies, for example tendencies and capacities and so on; but all bodies and things that have their existence in bodies are composite.

That bodies are composite is self-evident; but Theophrastus says that the capacities that have their existence in underlying bodies are also composite, and generally all forms in a substrate, firstly in virtue of being composed of genus and differentiae, and secondly, even if they are simple when considered according to their definitional specification, nevertheless the definitional specification has no existence in reality but only in thought. As things with substantial existence, on the other hand, they are considered with their underlying substrate; for sight that exists in actuality is not merely the faculty but includes the optical breath.[18]

Furthermore the natural philosophers,[19] when considering them as real and physical things, consider them with their substrate. For their definitional specification, as I said, has existence only in thought;[20] in this way Aristotle wanted there to be no transcendent Forms of them either, by not having the natural kinds subsisting in separation from bodies, but only existing in mere thought.

Furthermore nothing will distinguish these things <sc. things existing in a substrate> from the form of body. So just as the latter is simple according to its proper specification (I mean three-dimensionality)[21] but as regards its existence it needs matter as well – and so after all existing body is not simple but composed of matter and form – so also the same applies to these things; for when they are considered according to their proper specification they are simple, but when we consider them as existent and actually real, we shall be considering them with their underlying body. And thus they are not simple but composite.

If, therefore, physical things are either bodies or have their existence in a body, and these things are composite, then physical things are composite.[22] But all composite things have elements and causes and principles: for simple things are the elements of composite things. Hence physical things have principles and causes and elements. This is how Theophrastus supports the minor premiss. By adding to this the major premiss (that everything that has principles or causes or elements becomes known when they have become known) he thus draws the conclusion that physical things become known when their principles have become known.

Plato said that there were six principles of all things:[23] matter, form, efficient cause, paradigmatic cause, instrumental cause, final cause. He calls matter that 'in which' (for he calls it 'mother' and 'receptacle' and 'nurse'),[24] form he calls that 'which' (for this is what is peculiar to each of the things, rather than matter which is one and the same in all of them); efficient cause he calls that 'by which'; instrumental cause he calls that 'by means of which'; the paradigmatic cause he calls that 'with regard to which'; and the final cause he calls that 'for the sake of which'. Of these six things, he calls three 'auxiliary causes' (*sunaitia*),[25] and three 'causes' (*aitia*); the efficient, final, and paradigmatic causes he calls 'causes' because they are separate and in control of their own activity, but matter, form, and the instrument are 'auxiliary causes' because they are not in control of their own activity.

Plato, then, said that the principles of things were six, as I said, but Aristotle said four: matter, form, efficient cause, final cause. As a natural scientist he omitted the instrumental and paradigmatic causes, since nature does not create with regard to a paradigm: being a sort of life she creates in the manner of life, rather than as a sort of knowledge like the mind. As a natural scientist, then, Aristotle does not refer to the paradigmatic and instrumental causes; for it is possible to do physics theologically, as Plato did in the *Timaeus*, discussing causes separated from physical things, and also to do theology in the manner of physics, as Aristotle did in the *Metaphysics*, deriving his teaching on divine matters from physical things. This sort of scientist, then, – I mean the sort who discusses the causes immanent in the physical things – will not require the paradigmatic cause for the reason specified, nor the instrumental cause because he gives an account of all the natural and first causes in general. Had his account concerned one particular subject in nature, as perhaps with the medics concerned with human bodies, he might have been able to say that temperaments perhaps, or innate heat or something like that, was an instrument of nature; for nature carries out her own tasks by means of these things. But since he is seeking the causes even of the temperaments themselves, and of the primary elements (i.e. the four elements), reasonably enough he does not employ the instrumental cause. For nature does not construct the primary elements by means of some instrument, but she creates them directly out of matter and form.[26] Aristotle calls matter that 'from which', form that 'according to which' (for things are characterised and spoken of in accordance with the form), the efficient cause he calls that 'from which the origin of movement derives', and the final cause, like Plato, he calls that 'for the sake of which'.

<Preliminary comments on the phrase 'principles or causes or elements'>[27]

Either Aristotle is using 'principles' and 'causes' in parallel to mean the same thing, so that he is saying that both the efficient and the final are principles and causes, but he calls the other two (the material and the formal) 'elements'. Or, alternatively, (a) he predicates the <first> term, 'principle', of all of them in common; indeed he clearly calls each of them 'principle' all over the place: speaking of the principles of physics he says 'necessarily the principle must be one or more';[28] and again speaking of matter and form, 'that the principles must be opposites is clear indeed';[29] and again about the same things he says 'one principle, then, is this',[30] meaning matter, 'and another one is the specification …',[31] i.e. the form; and (b) he predicates <the second term>, 'causes', of the efficient and final, and 'elements' of the other two, matter and form.

He added 'that have principles' because not every science or discipline uses its own proper principles, but there are some that use the principles of the other sciences, such as dialectic which creeps into all the sciences: it tries to demonstrate medical theorems using medical principles, mathematical theorems using mathematical principles, and so on. Thus this discipline of dialectic does not have principles of its own, but engages with the various kinds of subject matter in this way, by using the principles belonging to others. The same goes for rhetoric, which always uses the principles of other sciences to support a proposition put forward; for this reason it is said to be the converse of dialectic, because both are involved in the same way with the same things. A 'discipline' (*methodos*) is a procedural methodology involving reason.[32]

<1.1.2 Textual analysis and exegesis, 184a10-16>

184a10 Since knowledge and understanding come about in all the disciplines …

We have spoken of the fact that Aristotle gave the major premiss of
7,1 the syllogism but omitted the minor premiss and the conclusion.[33] Hence the major premiss is placed right in the opening words. Because of this, in order for us to understand what is meant more clearly, we need to remove the conjunction 'since' and read the phrase as follows: 'knowledge and understanding come about, in all disciplines that have principles or causes or elements, from identifying these'. The general premiss is clearer thus. Or as follows: 'in all the disciplines that have principles or causes or elements, knowledge and understanding come about from identifying these'. Then, taking the conjunction 'since', we need to put it outside the minor premiss: 'since there are principles and causes and elements of physical things too, it follows that we shall come to know physical things when we have come to know their principles and causes and elements'. Then finally to read what follows from the conclusion, namely 'it is clear, therefore, that also for the science concerning nature …', and so on. So much concerning the syntax of the sentence.

Some people do not take 'knowledge' and 'understanding' as indicating the same thing, but rather take 'knowledge' (*eidenai*) for the simple cognition without demonstration, and 'understanding' (*epistasthai*) for that with demonstration. Others, however, take both in parallel as signifying the same thing. But even Plato brings in the term 'knowledge' (*eidenai*) for the most accurate and scientific cognition: for he says in the *Phaedo*, 'knowledge (*eidenai*) is this: having got hold of the understanding (*epistêmê*) of something, to retain it and not to have lost it'.[34]

184a11 that have principles or causes or elements ...

If we take the conjunction 'or' as disjunctive, the premiss becomes more general and the set has more members; for the disciplines that are not physical but are rational and intellective have principles and causes but do not have elements; but if we take the 'or' as subdisjunctive, in place of 'and', the argument will only allow for the physical sciences, which have principles *and* causes *and* elements. Only in composite things are there elements, which is the same as to say only in physical things. For an element is that which is immanent in the thing and becomes a part of the composite, but not a part like the parts <of the body> in the normal sense, i.e. the uniform and organic parts: these have their own circumscribed location and activity, and each of them is allotted a certain part of the composite; but the elements, by contrast, penetrate through and through each other and through the composite, so that there is not a jot of the composite that does not share in the elements.

The term 'principle' is more general (it is applied to form and to matter and to the efficient and final causes and the rest), but 'cause' and 'element' are more particular: 'cause' is applied to principles separate from the product – the efficient principle and the final principle and to the paradigmatic and the instrumental – and 'element' is applied to immanent principles – I mean form and matter, which become part of the product (but not part like those called non-uniform parts or the uniform parts,[35] but as complementary of the essence of the thing; we have already said in what respect a part differs from an element).[36]

184a12 For it is then that we think that we know each thing, when we have identified the first causes and the first principles and as far as the elements.

Aristotle says 'first' principles and causes, either (a) in order that we should begin from the very first principles and thus proceed down to the proximate ones (in fact we shall come to know each thing accurately when we know both the very first principles of it and the intermediate ones and the proximate and immanent ones, which is what he indicated by saying 'as far as the elements'; the proximate efficient cause of physical things – nature herself I mean – is immanent in the thing and is an element of it pervading the whole of it), or (b) by 'first principles' we are to understand first *of each thing*. This he also indicated by saying 'it is then' that we shall know 'each thing'. For example we shall know meteorological phenomena when we know their first principles *qua* meteorological things, knowing that they have this matter and that form, and this efficient cause, and the final cause in the same way. To know the very first matter is not to know the first principles simply *qua* meteorological, but simply *qua* bodies. So when we know the individuating efficient and final principles of each thing and have got as far as the knowledge of their elements – I mean their individual matter and form – then we shall in fact have accurate knowledge of each thing.

> **184a14** It is clear that also for natural science we should try to define [firstly the matters concerning the principles].

We said earlier[37] that Aristotle omits the conclusion of the syllogism, which was 'therefore physical objects become known when their principles have become known', and gives what follows from the conclusion. This is that if physical things become known when their principles have become known, it is clearly necessary for those who wish to come to know physical things to try to draw definitions concerning the principles of physics.

<1.2.1 Exposition and discussion: the appropriate method for physics, 184a15-b14>

> **184a16** Our route is naturally from things that are more knowable and clearer *to us*, towards those that are clearer and more knowable *in their nature*.

Having said in the earlier introduction[38] that we shall not acquire understanding concerning nature except by gaining knowledge of its principles (for physical objects have principles or causes or elements, and every discipline that has principles or causes or elements becomes known when the principles have become known; therefore physical objects become known once their principles have become known) now Aristotle wishes to say in what way the principles become known.

We need to be aware that Aristotle himself said, in the *Posterior Analytics*,[39] that there are two ways of acquiring scientific knowledge, (a) the demonstrative method, and (b) the didactic method, and that these are in opposition to each other. For the demonstrative method demonstrates secondary things from things that are first and more fundamental in nature, while the didactic method, although it is also a demonstrative approach, demonstrates things that are prior from things that are posterior in nature, using smaller proportions of demonstration;[40] this approach he also calls 'evidential'.[41] For example if someone, on seeing smoke, were to say that there was a fire there: he has argued for what is prior from what is in nature posterior.

We use this didactic method for certain things because we are unable to use the prior and more demonstrative method due to the fact that the nature of things is frequently in this condition, either through not possessing more fundamental principles, or due to the fact that, though they do possess them, those principles are less clear and less knowable to us. Thus Aristotle himself, in the *De Caelo*,[42] wishing to demonstrate what the shape of the moon is, argued *from* its phases *for* it being spherical: for it appears twice crescent-shaped, twice sliced in half, twice gibbous and once full; but these phases cannot possibly belong to any shape other than spherical. Here, then, he demonstrates evidentially what is prior in nature (the shape) from things that are posterior in nature (I mean the phases), when he ought to have shown first that it is made of the fifth element and because of this must be spherical (for it is impossible for there to be an eternal body that is not spherical) and then from the fact that the body of the moon is spherical to have shown that because of that it must have phases in this way.[43] This would follow more strictly the principle of arguing for posterior things from more fundamental things. But since the phases of the moon are more knowable to us than its shape, and its shape is more knowable to us than its substance, not unreasonably, he argues for the less clear things from the things that are clearer.

Also when explaining the more universal genera, Aristotle starts his explanation from things posterior in nature, giving descriptions and not definitions of them, or when he explains matter by a stripping-off process:[44] concerning these things it is necessary to start the explanation from things posterior in nature due to the fact that they do not possess more fundamental principles.

Since the present task is to get to know the principles of physical things, it is necessarily by the second of these methods that we must get to know them; it is, after all, impossible to do so by the first method. For if we wished always to establish our principles on the basis of principles, we should go on *ad infinitum*, always seeking further principles for the principles already adopted. So we are left with employing the second approach to the study of the principles, and progressing from things that are secondary in nature up to knowledge of these things. For it is always the case, Aristotle says, that things that are in nature prior and more clear are to us posterior and less clear, and things that to us are prior and more clear are in nature posterior and less clear.[45] Hence we have to begin our study from things that are more clear and prior for us, but in nature posterior and less clear. These, he says, are the 'indiscriminate things';[46] for they are more knowable to us and clearer, though less knowable in nature. The 'indiscriminate' are the universals,[47] so that we are to start our study from the universals. This is what Aristotle says.

It is a matter of debate in this connection, why Aristotle says that universals are posterior in nature and less clear, but to us prior and more clear, given that elsewhere he suggests the opposite, that knowledge of universals is less clear to us and naturally comes later;[48] for grasping the universal is the work of understanding (*epistêmê*) alone, and understanding comes later to us; and <it is a matter of debate> what things he means by 'universal' (*katholou*) here.

Our first answer is that by 'universal' he means here 'the particular' (*merikon*), which he has differentiated from the individual (*kath' hekasta*), as he says in the *De Interpretatione*,[49] in virtue of the fact that the individual is definite, whereas the particular is indeterminate. For the particular is nothing but an indeterminate individual capable of application to many things.[50] Not unreasonably Aristotle also called it 'indiscriminate' and 'universal', because of this very
11,1 property of being indeterminate: 'universal' because of its capacity to apply to many things, while the individual applies to only one; and 'indiscriminate' because it applies in an indeterminate and inarticulate manner to the things it does apply to. If we know the individuals in accordance with the universal, the knowledge we have of them is clear and articulated; for when I say that Socrates is animal or human, I have the nature of human and of animal defined and articulated; but when we know the individuals in accordance with the particular, as when we say that something is an animal or someone is a human being (for example that the one approaching is an animal or is a human being),[51] we know that in an inarticulate and indiscriminate way. For 'an animal' can be a human or a horse or any of the others, and similarly 'a human being' can be Plato or Alcibiades or someone else. So the first knowledge we get is knowledge of the indiscriminate things. For when we see Socrates from afar, we know first that it is an animal, and in this respect the knowledge is an indiscriminate knowledge of him as animal; for 'an animal' might be any one of the other animals. And then 'a human being' works the same way; even if it is distinguished from other animals, still we have a knowledge of it that is nonetheless indiscriminate: for it could be any one of the other human beings. But if we add to this indiscriminate knowledge the individual properties of Socrates, we make it the most precise knowledge of Socrates. So in this way, Aristotle says,[52] children start off by having an indiscriminate knowledge of fathers, and know them as human beings, not as fathers; hence they think all

human beings are fathers. But later when they can eventually attend to the properties of their father, then they get hold of an articulated knowledge of him and so distinguish him from other human beings.

Some people offer a different interpretation of these remarks. In order to preserve the proper meaning of 'universal', they say, we must say that Aristotle meant by 'universal' what is in the many things.[53] For what is before[54] the many is clear to us but known at a late stage, while what is upon[55] the many posterior things comes at the end of the process of reasoning, when we are, at length, able to get an impression, from the particular things, of the nature that belongs to them all in common. And besides, discussion of what is before the many is the task of the theologian above all, while discussion of what comes later is the task of the dialectician; but it befits the natural philosopher to discuss what is in the many. So it is this, they say, that is said to be prior for us and more clear and indiscriminate. For when we see someone coming, we identify him first in accordance with the indiscriminate universals; for the first thing we think is simply that it is body, then that it is animal, then that it is human, but not yet what sort of body or what sort of animal or what sort of human being.

In my own opinion this second interpretation does not harmonise either with Aristotelian teaching, or with the actual nature of things. Firstly the examples that they offer are not examples of the genus that is in the many, but of the indeterminate particular.[56] What else is the genus in the many if it is not the aggregate of all the particular things, or rather what is predicated of all the particular things in common? But when I say that the one approaching is animal or human, it is not that I think that the one approaching is the aggregate of all the animals, or that it is the animal that is predicated of all the particular ones, but I understand it to be the individual animal that is ranked immediately below the species. This is indeterminate, because I do not have any means of shoring up my intuition as to which of the individuals ranked below the species it is. For this reason I claim that 'an animal' is particular and not individual [because, being indeterminate, it can apply to many individuals, and in this sense can be more general than the individuals][57] in that the individual, being definite, cannot in any way apply to anything else, but this can apply to many things, because it is indeterminate.

So firstly, as I said,[58] the examples do not indicate the genus in the many, and secondly the genus in the many, even if it is in reality both posterior and less clear in the particular nature[59] (for nature knows the same thing as she makes, and nature does not make any of the universals but only the individuals, and then subsequently constructs the universals from the aggregate of the individuals), nevertheless it is still posterior for us in terms of knowledge, and for that reason also less clear. For we come to know the individual first, since it is of the individual that perception gets a hold, and we operate first of all by perception and subsequently by reason. It is reason that collates the particulars known to perception and calls the aggregate of all of them, or simply what is common to them all, the 'genus'. So that the knowledge of the genus that is in the many is posterior for us as well, and for this reason also less clear; for we bump into individuals more readily than we bump into universals.

Furthermore this interpretation does not harmonise with the Aristotelian examples either. Even if the child identifies every human being as father, yet he does not have an idea of the universal, but of the particular, the indeterminate particular; he does not have the characteristics of the individual articulated, so that he applies to everyone those characteristics of the human being that he picked up at the start. But as I said before,[60] we, when we identify the one coming as animal, do not identify it as a genus (for we know it is one thing), but as an indeterminate individual; and that is the particular. Hence the second interpretation is totally inappropriate.

We shall therefore understand the term 'universal' in accordance with the first interpretation,[61] as the particular which, in virtue of being indeterminate, is both indiscriminate and universal. For the genus is not indiscriminate; whenever I say 'animal' I have defined the substance and whether it is said of thousands or of one, the substance of the animal will not be any the more or less known; but whenever I say 'an animal', since the meaning conveys not only the notion of animal but also of the existence of some differentia belonging to it and separating it off from the others, but does not as yet add what the differentia is – for that belongs exclusively to the individual – for this reason it is indiscriminate and inarticulate. And as I said,[62] it is also posterior in nature; for this indeterminate thing is derived from the individuals, since it is a common term that can apply individually and properly to all the individuals. Therefore since things that are universal and indiscriminate in this way are more clear and prior for us, but it is necessary *** also in the discussion of the first principles to start from things that are indiscriminate and more general, but clearer to us.[63]

Someone might say that the things that are more general and indiscriminate are prior and more clear not merely to us but also in nature: for nature always begins her handiwork from things that are more general and inarticulate; for example if she wishes to make Socrates, first of all she lays down the specifications (*logoi*) simply of body, which is something general, not merely applicable to living things but to the inanimate too, and hence indiscriminate; for as regards the creation of the body it is unclear what sort of a body it is that has been presented; then as she proceeds she articulates it more and makes it animal, but this is still indeterminate and more general. Similarly by the same process out of the animal she makes human. And then finally adding the most characteristic differentiae she

makes the individuals, such as Socrates or Alcibiades. So that in nature too the things more general and indiscriminate are prior, and hence they are more clear in nature because more proximate and closer. In reply to this it is not reasonable to say that nature's aim is to make the individual and that it is in relation to the individual that she first presents the specifications of the more general things; for 14,1
knowing the individual is also what our knowledge has as its aim, and, as a sort of route towards that, it uses a starting point from what is more general.

To resolve this puzzle our reply is that nature never makes the things that are general and indiscriminate, but always makes what is individual and articulated. When we declare that the thing approaching is a body we can apply this notion 'a body' to anything – inanimate, animate, equine, human – and nothing prevents the mental image of 'a body' from being applied to any one of these; but it is not as if nature does the same, and when in the course of making Socrates she first puts forth a body, she puts forth a body of such a kind that the same one can do equally for a stone or a horse. Not a bit of it. Indeed the body will not even do for any human being you like, but rather a body that can only be Socrates' body – that is how nature puts it forth. Similarly she makes Socrates' animal and Socrates' human being. So that nature makes only particulars.[64] For even if she first of all puts forth the things that provide the material specification for the things that come after, and in this respect we have something in common with nature in that we begin from more material things, nevertheless we differ in that nature starts from particulars, and then makes the universal out of the collection of individuals, while we begin from the more universal and proceed to the more particular,[65] and nature always makes things that are articulated, and subsequently the things that are indiscriminate derive from these in the manner described, but we begin from the ones that are indiscriminate and proceed to the ones that are articulated.

But let us explain what is the aim of the project and the direction of the work as a whole, starting from the beginning. Our project, Aristotle says, is to get to know physical objects. 'Physical things' are clearly the things that have existence in reality and subsist, and these are the individuals; for none of the things that are more general and universal subsists in itself: what is not this stone or this plank in a full sense cannot be body in a straightforward way, and what is not this horse (e.g. Xanthos or Balios) or this man cannot be animal in a straightforward way. Nor does the universal come into being as a whole, nor does nature recognise it. For what nature knows, that she also creates, and what she creates, that she also recognises, and she creates only the particular.[66] So that is also the only thing she recognises. For if she had known the universal, clearly she would have made that.

15,1 Since our project is to get to know physical objects in reality, and everything becomes known when its principles have become known, so also we shall come to know physical objects in reality when we have come to know the proper principles of each of them. But since it is not possible to discover other principles for each of the first principles, he says we have to start from the things which are secondary in nature, but prior and more clear to us, and move up from them to knowledge of the principles. What are secondary in nature, but prior and more clear to us, are the things that are more general and applicable to many things, but indiscriminate, rather than the more peculiar and articulated principles of each thing. For the more general and inarticulate impressions always reach us first, and subsequently we make our way from those to the more peculiar and articulate ones (evidence for this would be knowledge in children and when we see someone approaching from afar) and more than everything else, knowledge naturally comes at its first impact from things that are more general.

For this reason, Aristotle says, it is from the things that are more general and indiscriminate that we too must set about the beginning of our study. Hence he begins by enquiring about the principles generally, whether they are one or many, and this is the study of things that are more general and indiscriminate; to enquire whether the principles of things are one or many is not to say anything about the distinctive character of each physical principle, but to study more generally how many physical principles there are of each thing, which is a subject posterior in nature, assembled out of things that are particular, and indiscriminate. For since nature, in making each of the particulars (given that nature makes none of the universals, as we said)[67] adopts some such principles for the creation of each thing, and adopts the same principles in the same way for all things, it is clear that this generality and universality is posterior, and derives its existence from the creation of the particular things; and that it is indiscriminate because it does not articulate what the distinctive character of each principle is.

Next after that the discussion moves on to something more articulated, but still inarticulate and indiscriminate as regards the peculiar principles of each thing. For when we have learnt that one of the principles is matter, another form, another privation, we have learnt a theory that is general and indiscriminate; for the matter that underlies heavenly and eternal things is different from the matter that underlies things involved in generation;[68] and, among the latter, the matter underlying things in the atmosphere is different from the matter underlying things on the ground; and the matter of plants is different from that of animals. So that even if he teaches us about matter in these books, nevertheless it is about matter under its more general description, matter that can be predicated of every <sort of> matter.

But the nature of matter is not unitary, not even of the prime matter. For that of heavenly things is different from that of things involved in generation.[69] However if one were schooled in the proper principle of each thing one would get a better vision of the fact that, in things, what underlies and is spread beneath, as in the case of matter, is a different thing from the form and shape that occurs in the matter (just as is the case with wooden or waxen products and such like), and that the same relation applies in all cases: that underlying the things in the atmosphere is the dry and moist vapour, and underlying the heavenly things is a fifth substance. For in fact one needs a lot of vision to grasp the peculiar principles of things. 16,1

Perhaps someone might be puzzled as to how, if Aristotle's discussion here is about the formless prime matter, we can say that he begins his study from things that are posterior in nature; for this one is the first of all principles. My reply is that in the first place, it is of bodies *qua* bodies that the very first matter is first principle in fact, but *qua* animals or plants or minerals or things arising in the atmosphere or heavenly bodies they have a different principle, a more peculiar one. So when we enquire into their first principle *qua* things in the atmosphere, we cannot say it is formless matter; that is no more the principle of things in the atmosphere than it is of all bodies as such. Instead we have to enquire into what underlies them peculiarly, their very first principle *qua* things in the atmosphere. Hence if his teaching is about formless matter, the study is nonetheless more general and indiscriminate; indiscriminate because the matter of heavenly bodies is no less formless than that of things involved in generation, and in this respect he has not said anything in particular about the peculiar matter of either the former or the latter, but has offered a general account about them; but this generality is clearly posterior in nature, since nature, as we said,[70] does not make any of the general things, but makes the individuals out of which the general thing is assembled.

Indeed nature does not make prime matter at all. It is, as I have frequently said, the individuals that come into being, and it is the proximate matter of each of these that the craftsman of each, nature, makes.[71] No nature makes prime matter; prime matter has been provided from the first, not by nature.[72] If therefore what are known to nature are the so-called 'first things' that nature makes, but the formless matter is not what nature makes, then when Aristotle's subject is formless matter he is not dealing with the things that are 17,1 first in nature and more knowable, unless one were to say that the formless matter is more knowable simply to the general nature that embraces all things.[73]

It is worth stopping to consider why what is said here is not in conflict with what is said in the first book of the *Posterior Analytics*.[74] Here when Aristotle says that 'our route is naturally from things that are more knowable and clearer to us to those that are clearer and more knowable in their nature',[75] he says that the universals are more knowable to us, while the individuals are less clear to us but more knowable in nature. 'Hence', he says, 'we must progress from the universals to the individuals; for the whole is more knowable by perception', he says, 'and the universal is a whole'.[76] That is what he says here, but he says the opposite in the first book of the *Posterior Analytics*; there he says that knowledge of the universals comes to us later: 'By prior and more knowable with regard to us I mean the things that are closer to perception', he says. 'But the universals are most distant by far, and the closest things are the individuals'.[77]

So we have two questions: (a) why he says here that the universals are more knowable to perception, and therefore that we have to start from universals and proceed to individuals, while in the *Posterior Analytics* he says the individuals are closer to perception, and the universals more distant; (b) why he says in the *Posterior Analytics* that the universals are primary in nature, while here he says the individuals are primary in nature. For even if it does not say so in so many words, nevertheless it follows of necessity: if the individuals are clearer in nature, evidently they are also primary. And we have just said that what nature makes, that she also knows best, and nature makes the individuals. Well the first of these puzzles we have virtually solved already, in that 'the universal' here means not the universal proper, but what is particular and indiscriminate because it applies to many things.[78] So the two do not contradict each other; for it is true that the individual is more knowable to perception – indeed perception does not grasp the universal at all – but since all nature proceeds from imperfection to perfection, for this reason it is in an indiscriminate and inarticulate manner that perception first hits upon the individual, given that it is not immediately capable of distinguishing the characteristic features of the individual from the rest. Hence it was for this reason that he called such perceptual knowledge knowledge 'of the universal', while in the *Posterior Analytics* he is referring to the universal proper; so that these are in agreement, not conflicting.

But how is it that here it is the articulate individuals that are said to be primary in nature, but there it is the universals? My view is that, just as in the case of individual animals it is not only in each part of the animal that there is a natural faculty that is entrusted with its management, but there is also one that governs the whole and effects a single animal – a harmony – out of all the parts, if the parts have to the whole the relation of matter, so also in the universe it is necessary that, just as there is the particular nature in each thing that is creative and preservative of that thing, likewise of necessity there must also be the one that harmonises the individuals into a single wholeness. Where else would the common properties in genera

and species come from, if there were no cause of that congruity? For the congruity is not spontaneous, nor by chance. So there is some natural cause for it, even if nothing of that sort subsists independently, but it inheres in the development of the individuals, in the way that the wholeness that applies to each individual, while not being separate from the parts, also inheres in the development of all these parts together. So that even if we say that none of the wholes occurs simply, that is true; for it does not occur itself by itself simply as a whole, but it makes its appearance along with the development of the parts. So in this way none of the universals occurs in itself, and yet it does make its appearance along with the things of which it is the universal as they occur. Since, therefore, each nature knows first of all that thing of which she is the cause, it follows that, just as that which creates the individual also knows the individual first and subsequently progresses from the individual to the universals, so in the reverse case the cause of the universal knows the universals first, and they are in fact clearer to her. In the physical treatise, then, since the aim is not merely to know things simply as physical, but also as such and such physical animals, such and such animals, such and such minerals, and in the *Meteorologica* as thunder and as lightning,[79] and in the same way for all the others, Aristotle not unreasonably says that the knowledge of the individuals comes to us subsequently, but the first is the knowledge of the more universal and indiscriminate things, according to the manner that we have already spoken of. In the *Posterior Analytics*, on the other hand, since the discussion concerns science, which is knowledge of universals, he says that the universals are primary in nature and more knowable, and that they reach us subsequently – and plainly he means there 19,1
that nature is the cause of the universals.[80] So that these passages are in agreement, not conflicting.

<1.2.2 Textual analysis and exegesis, 184a16-b14>

184a16 Our route is naturally from things that are more knowable and clearer *to us*, towards those that are clearer and more knowable *in their nature*.

Starting from the things that are clearer to us but less clear in their nature, Aristotle says, we end up at the things that are unclear to us, but clear in their nature. So since the route to knowledge of things in general naturally takes this form, he says, it is necessary that we should use this didactic method now too when we want to acquire knowledge of physical things.

184a22 It is from these, when we divide them, that subsequently the elements and the principles become known to us.

For by dividing the more general and confused we arrive at the more peculiar and articulated.

184a24 The whole is more knowable by perception, and the universal is a whole; for it incorporates many things as parts.

Just as in the case of perceptible things, if an animal were approaching, sight would more swiftly get a grasp of the whole body than it would of the head and the hand – hence one would say it was an animal sooner than <one would say> that it was a man that was approaching – in the same way in the case of things contemplated by reason, the things that are universal and general are clearer to us than the proximate and general. And indeed the universal is analogous in a way to the whole; for in the same way as the whole is inclusive of its own parts, so also the universal is inclusive of the particulars. But they differ from one another in that the universal gives its own name to those ranged under it, while the whole does not do so at all: the finger is not called a hand, nor is the nail called a finger. Given that the universal is more knowable to reason, just as the whole is more knowable to perception, and we have to begin from the things that are clearer and more knowable, here too, then, we must begin our study from these things. This is the way that the orators work, explaining the general headings first and then the peculiarities of each form. And this is the way that Aristotle himself, in the *Prior Analytics*,[81] puts the account of syllogism *simpliciter* first, before the account of the dialectical, demonstrative, and sophistic types of syllogism; for we observe the peculiar characteristics more easily from the common ones.

20,1 **184a26** Names undergo just the same, in a way, in relation to the specification: the name signifies a whole in an indefinite way.

The name signifies the substance of the thing in a rough and indiscriminate manner, for example 'human being', 'animal', 'circle'. The specification (*logos*), on the other hand, – that is the definition – signifies explicitly each of the properties that belong to the thing, and, as it were, divides it into its proper parts. For when I say 'human being' I have spoken a whole and indiscriminate utterance that incorporates in itself many things, while the definition resembles division into the parts: 'rational mortal animal', Aristotle says, signifies our underlying substance in a clear-cut way. And it was as well that he added 'in a way', because the way in which the name is what is divided by the definition is not the same as the way that the whole itself is divided into parts; rather it is what is signified by the name that the definition divides; and it does not happen that each of the things indicated by the definition is a part of the name, in the same manner as the individual is part of the universal. It is not the case that 'an', if it occurred as a syllable of '*anthrôpos*', would indicate animal, and another syllable would indicate rational, unless indeed it were in accordance with what is signified; for the parts of the definition are parts of what is signified by the name, while the present discussion concerns the utterance itself. In the case of the universal, however, e.g. animal, the individual animals are parts of it.

<2 On the principles of physics and the denial of plurality, Book 1 Chapter 2, 184b15-185a20>

<2.1.1 Exposition and discussion: a survey of Aristotle's Presocratic predecessors 184b15-25>

184b15 It is necessary that either there be one principle or more.

Here Aristotle launches into the discussion concerning the principles. He does, here too, what he always tends to do: he first refutes the false views of earlier thinkers. He collects the views of the earlier thinkers on the basis of a classification[82] that is scientific and at the same time lucid: scientific in that it is based on contradiction[83] (scientific classifications are based on contradiction, and they are scientific because they are inescapable; for nothing escapes the contradiction) and lucid in that it is <constructed> from things that are familiar to all: the one and the many are familiar to all. It is necessary, Aristotle says, that 21,1 the principles are either one or more; if such a phrase is not in itself a contradiction,[84] nevertheless it is equivalent to a contradiction; for what is not one necessarily must be more. 'It is necessary therefore', he says, 'that the principle either be one or more; and if one, either motionless or in motion, but if more, either finite or infinite'. Aristotle himself takes the classification this far at this point, but since he appears to add further sections to the classification in subsequent passages, it is a good idea for us to give the classification in full from the start.

It is necessary, therefore, that the principles be either one or more, and if one either motionless or in motion, and if in motion either finite or infinite, and again if motionless either finite or infinite, and if more, either finite or infinite, and if finite either in motion or motionless, and if infinite either in motion or motionless.

This is the classification of the principles; but if one were to wonder why we divided the one principle immediately into motionless and in motion, and then into finite and infinite, whereas the plurality we divided first into infinite and finite, and then into in motion and motionless, the answer we shall give is that it was possible to divide both sections of the classification in the same way, except that since motion is more closely related to the one than is the infinite, whereas the infinite is more closely related to plurality than is motion, for this reason we divided the one principle immediately into in motion and motionless, but the plurality into the finite and infinite.

That there was *one principle, motionless and finite*, was the opinion of Parmenides[85] and Xenophanes.[86] But it is abundantly clear, in fact, that their discourse was not about physical objects; for, as Aristotle himself said in the *Metaphysics*,[87] it is not as if these men were worse than crazy, that they should think that fire and water are no different. And it is plain from Aristotle's own words that these men were not natural philosophers; for when he says

> 'and if one, either motionless, as Parmenides says, and Melissus',

he adds

> 'or in motion, as the natural philosophers',

implying that Parmenides' school[88] were not natural philosophers.

Again it is clear from the things that Aristotle himself says in the
22,1 *De Generatione*[89] and what the commentators on that text say,[90] that they <sc. Parmenides and his school> were not talking about physical things. <Aristotle and the commentators say> that in the *Towards Opinion* (*pros doxan*) Parmenides said that fire and earth were the principle of all things, but in the *Towards Truth* (*pros alêtheian*) he said that the universe was one and finite and motionless. But we are not to suppose that in the *Towards Opinion* he was giving not his own opinions, but popular ones, whereas in the *Towards Truth* he was expressing his own opinions, for there too, even in the *Towards Opinion*, he wrote what he himself thought. But because his argument in the *Towards Truth* was concerned with the intelligibles, and these he called real being[91] and truly being,[92] on that account he gave the book that title. But in the *Towards Opinion* his argument concerned perceptibles, which Timaeus also called 'opinables':[93] 'for the eternal beings are accessible to intelligence with reason', Timaeus says, but the perceptible things, which he calls things that come to be and perish, 'are opinable to opinion with unreasoning perception', he says.[94] Thus it was about the intelligibles that Parmenides said the stuff about them being one and motionless and finite. Since the intelligibles are more related to the one, in virtue of being closer to the one principle of all things, that was why he called them 'one'.[95]

Parmenides and Xenophanes said that the principle is one, finite and motionless; but Melissus, who was also talking about the same things, as I have already said, also held that it was one and motionless;[96] except that Melissus said that the one was infinite, while

Parmenides' school said it was finite, and this was because they were attending to the definitive and form-giving power of the intelligibles, while Melissus was attending to the infinitude of the power.

Those who said straightforwardly that being is one and at the same time motionless were thinking on the right lines; for if it moved there would no longer be one: for then there would also be motion itself and a place or time or something of that sort in which the motion took place. But we need to be aware that these people who said that all is one were not giving a discourse concerning the principles. If they were saying that the *principle* of things that are is one, they could no longer have said that being is one, for once the principle exists so too must that which is derivative from the principle. These are in the category of relatives, for the principle is principle *of* some things. So their discourse was not about the principles but simply about the things that are.

But if their discourse was not about the principles, why does Aristotle criticise these men on the grounds that they were mistakenly suggesting that the principle was one? My reply is that, even if they were not talking about the principles, nevertheless Aristotle criticises the argument as if someone had been suggesting that the principle of physical things was one, because he wants to demolish such a theory.

We have spoken of the ones who said that the principle was one 23,1
and motionless; among those who said that it was *one and in motion*, some said it was finite, others that it was infinite. The following said it was *one, in motion and finite*: Hippasus,[97] Heraclitus,[98] Thales,[99] Hippon (nicknamed atheist).[100] Heraclitus and Hippasus said that fire was the principle of things because it is made of finer particles than the others and is easily moulded. (We are not to think, on account of the description of its form, that it relates to the other things as something that encompasses the rest; the present discussion concerns the material principle.) Thales and Hippon said water, because of what is reproductive and because they observed that seminal fluid is moist and that seeds, even if they are dry, do not grow unless they are first moistened and then only once they are sodden.

One principle that is in motion and infinite was posited by those who posited air and those who posited what is between either air and water, or fire and air. Thus Diogenes of Apollonia[101] and Anaximenes[102] proposed that air is the principle, because it is both easily moulded, and like something bodiless and rather imperceptible within bodies. Anaximander[103] posited what is between air and water or between fire and air: for since the transfer is not immediate in the change from water to air, but first of all the water is sort of vaporised and subsequently the vapour thins and becomes gaseous, and similarly for the change from air into fire, Anaximander said this intermediate is the principle, a stuff which is more dense than air but more rare than water, or more dense than fire and more rare than air. Anaximander was more correct than the others, in that he did not say that matter was some one of the elements, but a different thing besides them; but he erred in that he posited something endued with form in saying that it was more dense than air but more rare than water.[104]

These thinkers said that the air or the intermediate was infinite,[105] since they observed that if things were always coming to be out of them, but they were not infinite, the process of coming into being would necessarily come to a stop.[106] For they did not pay attention to the <notion of> exchange and the <way> things change into one another.[107] So these were the people who said there was one principle, and that is how many said it; none of the people who said there was one principle proposed that earth alone was a principle, due to its resistance to movement or change.

Among those who posited *a plurality of principles*, some posited an infinite number, some a finite number, and in both these groups there were some who held that they were motionless, and others who held that they were in motion. In the first place Timaeus posited *several, both finite and motionless*, namely god, form, and matter,[108] while Empedocles had *several, finite and in motion*, the four elements and strife and love.[109] But Empedocles held that there were two worlds:[110] (a) the world of the four elements which is the perceptible world and (b) the Sphere,[111] i.e. the intelligible world (which he calls the Sphere because it is turned towards itself, and is closer to the one); and he said that these two worlds change into one another. When love is predominant the elements change into the Sphere, but when Strife is predominant the Sphere changes into the elements. By this he does not mean that the worlds change into one another, but rather he refers to the transfer of our soul into these worlds. In our soul there is both a differentiating faculty and an assimilating faculty; when it functions according to the assimilating faculty, which Plato called the 'circle of the same',[112] it comes closer to the intelligibles. Empedocles called this faculty 'love' because love is unifying. On the other hand when it functions according to the differentiating faculty, which Plato called the 'circle of the different',[113] but Empedocles called 'Strife', it comes closer to perceptible things. Empedocles spoke of the soul's change between the perceptible and intelligible worlds as a change of the worlds; hence his much-quoted saying about the soul:

> So I am here, a fugitive from god and a wanderer,
> placing my trust in raging strife,[114]

that is having relied on the differentiating faculty of the soul.

Thinkers who posited *several, infinite and in motion* were Anaxagoras, Democritus, Epicurus,[115] and Leucippus.[116] Anaxagoras[117] posited the homoiomeries[118] as principles of things; a homoiomery is that of which the part is similar to the whole.[119] So the homoiomeries exist in each other; for example in wood, which is a homoiomery, there exist flesh and bone and water and fire and gold and altogether everything, and in each and every thing there are unlimited homoiomeries. He posited this theory because he saw that all things come into being from each other, if not by a direct change, at least through a number of intermediaries. So he imagined a kind of matter that was fit to be transformed into everything, but he did not have the ability to work it out completely; he suggested 'all things were together', and that in this way the coming into being from one another came about by a process of separation and not by a change of the substance. He did well to suggest that the uniform parts were in one another, but the non-uniform ones were not; for no one would suppose there is a hand in a face or a foot in a hand.

Democritus and Leucippus and Epicurus posited atoms and the void, and that both the void and the atoms that are in it are infinite. The term 'atoms' refers to certain bodies that are invisible due to their smallness and indivisible due to their hardness, like the dust particles that appear in the sunbeams that stream through the windows, which become invisible when the beam is not shining on them, not because they are not there but due to their smallness.

No one suggested that the principles were *several, infinite and motionless*.[120]

<2.1.2 Textual analysis and exegesis, 184b15-25>

> **184b20** And if infinite, either in this way as Democritus had them, one in genus,[121] but differing in shape or form, [or even opposite].

'Genus' means what is underlying and the substance of the atoms. For Democritus said all the atoms were of one substance, but they differed from each other in their shapes, in that some are spherical, some are cubic or pyramidal, or have one of the other shapes. In the *Posterior Analytics* Aristotle frequently uses 'genus' as a name for what is underlying in this way.[122]

184b21 In shape or form.

Aristotle is saying the same thing with two words in parallel.[123] For Democritus said that in the atoms the shape has the function of form.[124]

184b22 Or even opposite.

Either (a) this says that Democritus thought that the atoms were one in kind, but that they differed in respect of their shapes, and in fact that they did not just differ but were even opposite. For Democritus said that heat and chill and white and black were not in the atoms, but that these effects were produced from their shapes and the condition of the atoms relative to ourselves; for the spherical ones, being mobile, are causes of heat and fire – since they are mobile they penetrate and pass through quickly, and this cutting capacity and mobility is a characteristic of fire; whereas the cubic ones, say, being thrusting and pressurising, produce chilling. For the cold is associated with compression. And he says it happens in a similar way for the colours. When the vertices of, say, the pyramids strike one's sight, it makes the impression of such and such a colour, for example white. For white is something that is piercing to the sight, and the sour is also penetrating, and that is what the vertex of the pyramid is also like. But when the bases <of the pyramids strike one's sight,> it gives the impression of black, for black is compacting, and what is blunt is like that; for it compresses, and presses together things that are separated into the same place by its compression. So since the atoms produce opposite effects by their different shapes, he says not only that they differ in their shapes, but also that they are opposite.

Or preferably (b) this phrase applies to Anaxagoras, who said that there were also opposite homoiomeries, for example of fire and water. This is closer to the truth, to preserve the classification by division as well.[125] For having said 'and if infinite, either in this way as Democritus had them, one in genus, but differing in shape or form' he then adds the second limb of the classification by saying 'or opposite' as if he had said 'or, as Anaxagoras says, opposite'.

> **184b22** Those who ask how many things there are make their enquiries in a similar way.

Having set out his classification, Aristotle now wishes to indicate that he has not been doing something superfluous or unprecedented in using this classification concerning the principles, but that all his predecessors in natural science had used the same classification in their discussion of the principles, when they enquired whether the element from which everything derives is one thing or more, and whether they are in motion or motionless.[126]

<2.2.1 Exposition and discussion: the proper task of physics, 184b25-185a20>

> **184b25** Now investigating whether being is one and motionless is not an investigation into nature.

Aristotle has set out the opinions of the natural philosophers, concerning principles, as a classification, and has indicated which of the earlier thinkers was prominent in each limb of the classification. Now he considers which of the opinions are subjects about which it is the business of the physicist to engage in discussion, and which are not. For the expert has the task of considering the nature of the subjects brought to his attention, and discerning which of them are appropriate for him and which are not. A carpenter, for instance, when faced with a great variety of pieces of wood, will select some as suitable for his work and reject others. This is why Aristotle now picks out one limb of the classification (the one that says that the principle is one and motionless, in which the school of Parmenides and Melissus were prominent) and says that it is not the business of the natural philosopher to discuss with these people who deny the physical principles. For it is not the business of any of the particular sciences to debate with one who denies the principles belonging to that science, since it is also not the task of a particular science to demonstrate the principles proper to the particular sciences. That is the task of the first, unhypothetical, philosophy;[127] it demonstrates the principles of all of them, while the particular sciences take their own principles as hypothetical assumptions. Hence debating with one who denies the principles of any science whatsoever is not the business of that science, but either of the science next above it, or of 'the common science of all'.[128]

By the common science of all Aristotle means either the first unhypothetical philosophy, or dialectic. Dialectic also demonstrates the principles of all of them. This is how he defines it in the *Topics*:

> The aim of the subject is to find a method whereby we may be able to reason logically about everything that is presented by established opinions.[129]

Dialectic differs from the first philosophy in that the latter constructs its proofs from self-evident premisses and common notions, whereas dialectic works from established opinions.

So the geometer will not engage in discussion with one who denies that a point has no parts, or that a line is length with no breadth;[130] he will hand over to the metaphysician to prove those things.[131] Similarly the doctor will hand over to the natural philosopher to prove that the body is composed of the four elements and <to provide proofs> concerning those elements themselves; for the natural philosopher is naturally ranked before the doctor. And again geometry will prove the principles of optics, music will prove those of grammar,[132] and arithmetic will prove those of music. The expert in grammar takes certain letters as long and certain letters as short, and the same letters as now long and at other times short. He does not himself know the reason, but the expert in music demonstrates it. Again the expert in music accepts that this string, in relation to that string, has the ratio three to two, or two to one, or four to three, taking the relations of these numbers solely in strings. But the expert in arithmetic examines in general what the ratio of two to one is, and the other ratios, whether it be in strings or anywhere you like.

28,1 They say that arithmetic takes precedence over even geometry itself – and evidently that <it takes precedence> over every mathematical science;[133] for the geometer takes a side that is double another, or one third more, and to this extent he is taking the double (or what have you) in size. But the expert in arithmetic takes these things in general and shows what the ratio of the double, and of the others, is. But the metaphysician demonstrates the principles of all <the sciences>.[134]

For these reasons it is not the business of a man of physics to engage in discussion with Parmenides and Melissus, who deny the principles of physics. When they say that being is one and motionless, they are denying the principles by suggesting that being is one, and they are denying that they are physical principles by suggesting that it is motionless.

For when they say that being is one,[135] either they are suggesting that it is the principle or that it is derived from the principle. But if they were suggesting that it is the principle, there is every necessity that there be things derived from the principle. These are correlative things: for 'the principle is a principle of something or some things', as Aristotle says.[136] So being is not one but several: the principle and the things derived from it. Hence they could not call the one a *principle*. Yet, again, if they suggested it was derivative from the principle, there is every necessity that they should posit its principle too; so that again being is no longer one. But in fact they posit one.

So they are eliminating the principle from both sides. For if it is a principle, being cannot be one. So in this way they are eliminating the principle. But they eliminate the principle *qua* physical by suggesting that it is motionless. For motion[137] is evidently a property of physical things: birth[138] and decay, growth, diminution, alteration, and change of place occur in physical things. When Aristotle defines nature he says that it is a principle of motion and rest. So if one eliminates motion one eliminates the physical principles. This is the reason that he says it is not the business of the natural philosopher to debate with these people; but it is as natural philosophers that we are embarking on this subject.

Nevertheless, Aristotle says, since their enquiry has a philosophical character (for they do investigate things, though not as natural scientists, but as philosophers) we shall be taking issue with them. In particular, given that they seem to raise some physical puzzles, a reply on our part seems reasonable in this respect too. For in denying nature, by suggesting that being is one and motionless, they would be puzzling about nature.

<2.2.2 Textual analysis and exegesis, 184b25-185a20>[139]

> **185a3** For there is no principle any more if there is only one and one like that. For the principle is principle of something or some things. 29,1

Aristotle proves in the second figure that the people who suggest that being is one eliminate principles.[140] It goes like this: if being is one, there is no plurality; if there is a principle, there is a plurality; so if being is one, there is no principle. For if there is a principle, there must necessarily be the things derivative from the principle, but in that case there is necessarily a plurality.

'If one like that'[141] stands for 'motionless'. For given that it is motionless, there will be nothing derived from it, nor will it be derived from anything else; but if there is a principle there must be motion. For if it were in every way unchangeable and motionless, nothing would come into being from it.

> **185a5** Inquiring into whether it is one like that is the same kind of thing as debating with any other thesis put forward for the sake of argument [(such as that of Heraclitus, or if someone were to say that being is one person)].

'A "thesis",' as Aristotle himself says in the *Topics*, 'is a paradoxical view of someone well known in philosophy'.[142] For example Heraclitus said that opposites are the same; and similarly these, who say that everything is one and motionless, or as Zeno[143] said, that there is no motion, or as Anaxagoras said, that everything undergoes change.[144] So enquiring whether being is one and motionless, he says, is like debating with any other thesis, such as that of Heraclitus, who said that opposites are the same. Just as it is pointless to argue against such a thesis, or against one who says that one individual person is all that there is,[145] given that self-evidence is more powerful than any argument, so the same goes for the theory of Parmenides and Melissus.

'For the sake of argument' stands for 'for the sake of saying something', where they are saying it simply by asserting it, not in logical argument; as for example if someone were to say 'man is three-footed'.

> **185a7** Or solving an eristic argument.

The bit from 'enquiring into whether it is one like that' and so on, and 'or solving an eristic argument' refer to the same thing. An eristic argument differs from a sophistic argument in that a sophistic argument employs true material[146] but has an invalid form, whereas an eristic argument slips up in both respects, both in the material and in the form.

> **185a11** Which is true of both the arguments, both that of Melissus and that of Parmenides. [For they both adopt false premisses and are fallacious; or rather Melissus' argument is crude and poses no difficulty. But once one absurd premiss is granted, the rest follow. But this is hardly difficult.]

Both Parmenides' argument and that of Melissus are eristic, Aristotle is saying; for each of them uses false premisses and proceeds fallaciously. What false premisses they adopt and how they combine them fallaciously, we shall indicate by turning to the arguments
30,1 themselves. But even if both the arguments are eristic, Aristotle says, yet rather Melissus' argument is crude, and provides no philosophical puzzle. For Parmenides' argument, although it adopts false premisses and is fallacious, still has a certain subtlety and acumen; whereas Melissus' argument is superficial, and hence it is 'crude' and raises no philosophical puzzle for those who study it. 'But once one absurd premiss is granted', he can infer the rest in this way. 'But this is hardly difficult', Aristotle says; by this <he means> either, (a) establishing what follows once you have granted one absurd premiss (e.g. given the premiss that the earth flies,[147] it will follow that we are aloft, that heavy things come to rest off-centre and lots of other things); or, (b) it is hardly difficult to explain and refute such superficial sayings.

> **185a12** Let us take this as basic: natural things, either all of them or some of them, are subject to change.[148]

Let us adopt this as a hypothesis, Aristotle is saying. For it is not the business of the natural philosopher to provide proof about first principles, to the effect that natural objects, all or some of them, are subject to change. And indeed it is from perception that we get our belief that either all or some of them are subject to change. Aristotle has added 'some of them' because potentialities that are in a substrate, and irrational souls,[149] are natural things but are not subject to change. For they neither alter, nor grow, nor change location.

> **185a14** But then again nor is it appropriate to resolve all of them, but only those errors that someone has made in deduction from the first principles.

Because he has adopted as a hypothesis the view that natural objects are subject to change *before* refuting those <philosophers'> views, he defends this by adding the statement 'nor is it appropriate to resolve all of them', but <only> those errors that someone derives from the first principles while retaining the first principles. It is not necessary to enter debate with someone who denies the first principles and brings difficulties against the first principles that undermine them; nor is it necessary to resolve the puzzles adduced by that person. 'But then again nor is it appropriate ...' stands for 'for it is not appropriate ...'.[150]

> **185a16** For example it is the business of the geometer to refute the quadrature of the circle that is done by means of segments, but not the one by Antiphon. 31,1

Hippocrates[151] was a merchant of Chios, who was a victim of piracy and lost everything. He travelled to Athens to bring a case against the pirates and while he was staying in Athens, for a long time on account of the court case, he attended philosophy classes, and reached such a high standard in geometry that he attempted to discover the quadrature of the circle.[152] In fact he did not discover that; having squared the lune, he thought, wrongly, that he could go on from there to square the circle. For he thought that one could also deduce the quadrature of the circle from the quadrature of the lune.

The other man, Antiphon,[153] also attempted to square the circle, but without preserving the first principles of geometry. This is how he tried to do it: 'Suppose I construct a circle', he says, 'and inscribe a square in it;[154] then I divide in half the segments of the circle resulting from the square; then I draw straight lines in either direction from this division to the extremities of the segments, and make an octagonal figure. Now suppose that again we divide the segments that encompass the angles each in half, and again draw straight lines from these divisions to the extremities of the segments in either direction, we shall make a polygonal figure. Suppose, therefore, that we do this over and over, the result will be a figure of a great many angles, having angles that are extremely slight,[155] whose encompassing straight sides coincide with the circle on account of being so small. Granted that any given rectilinear figure can be squared, if I square this polygon, since it coincides with the circle I shall have squared the circle as well'.[156]

Hence Antiphon denies the principles of geometry; for it is a geometrical principle that a straight line never coincides with an arc of a circle,[157] but Antiphon allows that, due to smallness, a certain straight line coincides with a certain arc.

So Hippocrates, setting out from geometrical principles, and having squared some lunate segment of the circle, drew his next conclusion wrongly, in that he wanted to deduce the quadrature of the

circle from this as well. Whereas Antiphon drew his next conclusion by denying the principles of geometry, that a straight line never coincides with an arc. So Aristotle says that, as regards proving that the quadrature of the circle is invalid, it *is* the business of the
32,1 geometrician to refute Hippocrates' quadrature of the circle, since Hippocrates preserves the principles of geometry; but the geometer will not go on to refute Antiphon's, because he derives his conclusion by denying the principles of geometry.

> **185a17** However though they are not writing on nature, yet since it happens that they raise puzzles concerning physics, [perhaps it is as well to discuss them a little].[158]

Even if these men are not saying anything 'on nature'[159] (for it was not by employing the principles of physics that they argued that being was one and motionless), nevertheless since it happens that they raise puzzles concerning physics (for in suggesting that being is one and motionless they are denying nature, but it happens that in denying nature they raise some puzzles concerning physics), Aristotle says that it is necessary to debate with them a little – not on the grounds that their doctrine demands discussion or debate, but because of the reputation of the personalities. For even things that are abundantly clear often seem doubtful to ordinary people due to the authority of the personalities. And, in any case, the inquiry concerning these matters itself possesses a certain philosophy in its own right;[160] in other words, the pure investigation into the things that exist, whether they are one or many, would also fit into the first and unhypothetical philosophy.[161]

<3 Puzzles concerning being and the one, Book 1 Chapter 2, 185a20-186a3>

<3.1.1 Exposition and discussion: the structure of Aristotle's argument, 185a20-b5>

> **185a20** But the most appropriate starting point of all [is to ask how, given that being is said in several ways, they mean it when they say that all things are one, whether all are substance or quantities or qualities ...]

In what precedes Aristotle has distinguished whom the natural philosopher is to enter discussion with, and whom not, and he has said that it is not the business of the natural philosopher to enter into discussion with the followers of Parmenides and Melissus, because they deny the principles of physics in asserting that being is one and motionless. He then said that even if it is not the business of the natural philosopher to enter into discussion with these people, yet

<since> the inquiry has some kind of philosophy to it, it would be no bad thing to debate with them a little.[162] And besides, even if they do not have anything to say on nature, they do provoke puzzles concern- 33,1
ing physics in denying the principles of physics. So having said these things in the earlier passages, he now embarks from henceforth on the arguments addressed to them.

But before he addresses himself to what they said, he explores the question in its own right generally. Addressing oneself to their actual arguments, by which they attempted to establish their doctrine, is specific to the task of challenging them, whereas exploring the question in its own right generally, and demonstrating whether it can be so or not, belongs to the general inquiry concerning that question. When he has carried out that demonstration, he will then address himself specifically to their arguments. For we have to explore the questions in their own right as well, and refute the difficulties brought against them, lest they hang around to trouble the more simple-minded.

Aristotle employs a classification, saying how many ways we can speak of 'one', and then shows that it is not possible to say that being is one on any meaning of 'one'. What is one, he is saying, is one either (a) *nominally* or (b) *really*.[163] I say 'nominally one', as for example if someone were to say that the real person and the person in the picture were one,[164] because they have one name, 'person';[165] or if someone were to say that all the categories were one, because they have a common name, the name of 'being'. So what is one is one either (a) *nominally* or (b) *really*. And (b) if it is *really* one, it is in respect of either (b 1) *the universal* or (b 2) *the particular*; and if it is in respect of (b 1) *the universal* it is either (b 1.1) one in *genus* or (b 1.2) one in *species*:[166] (b 1.1) one in *genus* as, for example, if someone were to say all the particular substances were one, because they have a common genus, substance; or (b 1.2) one in *species* as if someone were to say that whitening and blackening were one, because they have a common name, the name 'alteration',[167] or rather the whitening in white lead make-up and the whitening in a woollen garment, because they have a common form, whitening without the qualifications;[168] or if someone were to say that all rational things were one, because the rational is a species of animal. If, on the other hand, it were (b 2) *really* one in virtue of *the particular*,[169] either it will be one (b 2.1) *as a continuous thing*, as we say that the plank is one, or (b 2.2) *as an indivisible thing*, like the point and the monad, or (b 2.3) *as what is one in specification*, like things that have several names, such as clothing and garment,[170] and arms and weapons;[171] for these are one in specification. So Aristotle takes the second arm of the classification, that of what is (b) *really* one, and, in this case, (b 1) *as universal*, and he says that those things which are said to be one *in respect of the universal* are evidently either *substance* or one of the *attributes*,

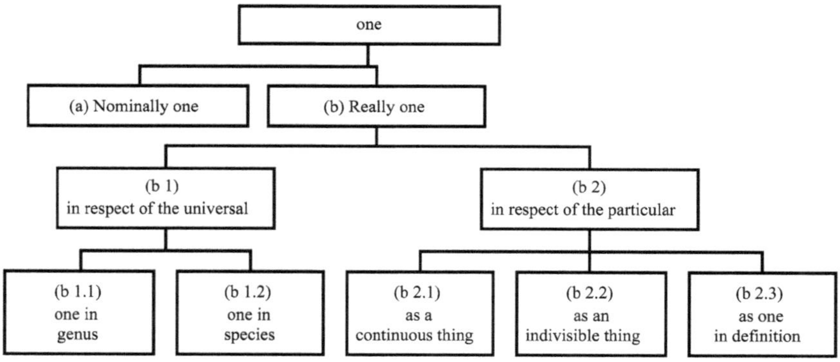

Classification of the senses of 'one'

quantity or quality or some other. Then he postpones *substance* to later[172] (for he is going to show that being cannot be one thing as substance either; for that way we shall be destroying the nature of the attributes), but for now he explores whether perhaps being is one thing *as a universal* in the range of *attributes*, such as quantity: for in virtue of all things being quantities, being is thus said to be one thing, quantity, according to the universal. But before he tests this, he inserts in between[173] the claim that being cannot be said to be (a) *nominally* one either; such as that several things, substance and attributes, 'are', and in virtue of the fact that they are all said to be 'beings' thus being is said to be one. But if this is so, he says, plainly what there are are several and not one. For, he says, one would not say that things that are *really* several are one, because of sharing a common name.

This is what he is asking, then, whether it is possible that all things are quantities, and thus all are said to be one (b 1.1) *due to their genus*. Well either there is substance in addition to this quantity, or there is not, he says; and if there is not, quantity will be self-supporting, which is impossible (for attributes have their being in substance); but if there is, either there is one and the same meaning for quantity and for substance and there is one thing signified by different words, or there is one meaning for substance and another for quantity. If therefore it is one and the same substance of quantity and of substance, they will be spinning together things that cannot be spun together by bringing substance together with attribute; but if substance is something else apart from quantity, again there are several beings.

Aristotle goes on, after saying these things, to show that his suggestion that they said that being was a quantity was not arbitrary. For Melissus says that this being is also infinite, Aristotle says,[174] and the infinite is either a species of quantity or a part of quantity, or indeed rather one of the *per se* properties. For one can speak of it as a species of quantity, saying 'of quantity one sort is infinite, the other is finite'; but more <accurately> it would be one of the *per se* properties. For the infinite is not a quantity in a straightforward way, but belongs to quantity.

There are two ways in which something can be '*per se*', as Aristotle himself says in the *Posterior Analytics*:[175] (a) that which has its subject[176] incorporated in the definition, as nose is incorporated in the definition of snub; for snubness is concavity in a nose, and snubness is a property *per se* of nose. Similarly we incorporate the subject in the definition of the odd and the even; for we say that the even is a number that divides in half. (b) And again hinnible belongs to horse *per se*; for it is incorporated in the definition of the subject. For a horse is an irrational hinnible animal, we say.[177] And it is a property of the circle that its radii are equal, since that is incorporated in the definition of circle. For it is a plane figure enclosed by one line, with one of its internal points such that all the lines drawn from that point to meet the line[178] are equal to each other.

So also the infinite belongs to quantity *per se*, because it incorporates the subject in its definition – I mean quantity; for we say the infinite is a quantity that cannot be traversed.[179]

Thus if Melissus said that being is infinite, and the infinite belongs to quantity *per se*, then he said that being was a quantity. For even if something else is said to be infinite, it is said <to be infinite> *per accidens* in virtue of participating in quantity; just as if white or black is said to be extensive, it is not said <to be extensive> *per se*, but *per accidens* in virtue of the surface, in which the colour is, being extensive, as Aristotle says in the *Categories*.[180] So even if substance is said to be infinite, it is not infinite *per se*; for it does not fall under any of the definitions of *per se*. For substance is not incorporated in the definition of the infinite, nor is the infinite incorporated in the definition of substance. So substance is not infinite *per se*, but *per accidens* in virtue of participating in quantity, to which the infinite belongs *per se*.

If, therefore, Melissus said that being is infinite, but infinite *per se* and not *per accidens* (because then being would no longer be one, if it is accidental to something), clearly he thought that it is quantity. So our suggestion that they said that being was quantity was not arbitrary.

<3.1.2 Textual analysis and exegesis, 185a20-b5>

> **185a20** But the most appropriate starting point of all is to ask how, given that being is said in several ways, they mean it when they say that all things are one.

Since being is said in several ways, Aristotle says, the most appropriate starting point would be to set out the senses of being and ask them, 'Which of the senses of being are you using?'. And again, since one is said in several ways, in what sense they are saying that this being is one. In fact the most appropriate starting point for any question whatsoever is to discover the proper meanings of the expressions used, and distinguish the issue in question from other things expressed using the same words ambiguously. For ambiguity is the cause of most error for people.

Aristotle interweaves the two classifications, that of one and that of being; he divides being into (a) substance and (b) attributes, and then immediately into (i) genus, (ii) species, and (iii) individuals. But what comes next, 'and yet since the one itself is also said in several ways',[181] is not straightforwardly a classification of the one in general, but of what is most properly one, that is the individual; he divides this into (i) what is continuous, (ii) what is indivisible, and (iii) what is one in specification.[182]

> **185a22** whether all these as substance or quantities or qualities.[183]

Some people[184] have understood Aristotle to be setting out the universal section of the classification[185] in these words, and that by saying

> 'Whether all as substance or certain quantities or qualities'

he is referring to what is one in genus,[186] whereas when he says

> 'And again whether substance, but all one substance, such as one person or one horse'[187]

he is referring to what is one in respect of species;[188] for they say that he sets out the particular section[189] a bit later when he says,

> 'What is said to be one is (a) what is continuous or (b) what is indivisible or (c) what has the same specification.'[190]

On the other hand our view is that, given that as soon as we hear 'one person or one horse' the mind is immediately directed to the particular, for this reason we say that in the phrase,

> 'Whether all as substance or certain quantities or qualities'

Aristotle is setting out the universal in general, both what is one in genus and what is one in respect of species, and in the phrase,

> 'such as one person or one horse'

he is setting out the particular. Afterwards he divides the particular itself, as I have already said.[191]

> **185a26** For all these differ greatly and all are impossible to assert.

The things just enumerated, Aristotle says, (that is substance, in my view, and attribute, and in the case of substance both the universal and the particular, and similarly quantity and quality and the other attributes) differ considerably from each other; and for each of them, if the followers of Parmenides are assumed to be speaking of that, impossibilities will always follow on that assumption.

> **185a27** For if there is to be both substance and quality and quantity, [and whether these are detached from each other or not, then there are several beings].

Here, as I have already remarked,[192] Aristotle inserts the suggestion that being is *nominally* one, in the way in which the ten categories are one in respect of the word 'being' itself, and have unity only nominally.[193] So Aristotle is saying that if they <sc. the followers of Parmenides> mean that being (which they are suggesting is also one) is itself both substance and also quantity and the other categories, whether these are assumed (a) to be detached and separate from each other or (b) to be in each other (as is in fact the truth, that the other categories have their being in substance), either way it follows that beings are several by nature, and that they have only the name in common, in that all are called 'beings'. Hence the unity is only nominal, not real.

> **185a29** But if they are all quality or quantity, [whether or not there is substance it is absurd].

Aristotle returns to the other arm of the classification, that suggests <that being is> (b) *really* one, but really one either as an attribute or as substance;[194] but now, as I said,[195] he postpones the theory about substance, and explores the one about attribute instead. So he says, if this being were one of the attributes, such as quantity, whether they were supposing that substance coexisted with it or not, the theory would be absurd. For if substance coexisted, then being would be no longer one but several; whereas if it did not coexist, but quantity subsists in its own right, the attribute will be separate from substance, which is not only absurd but also impossible. That is why, when he says 'absurd', he adds 'if it is right to call what is impossible "absurd" '.[196] But 'impossible' is a stronger form of 'absurd'.[197] For the contrary of 'absurd' is 'plausible'; for the absurd is what one would never guess because it bears no relation to any reason or argument. So 'plausible' is the contrary of absurd, and the contrary of 'impossible' is 'necessary'. But 'impossible' is a stronger form of 'absurd'; for what is absurd might perhaps also occasionally happen. For example no one would envisage the entire population taking a bath; to envisage that is false, but not impossible. But what is impossible is both false and impossible, such as if someone were to say that humankind has wings. That is false in a strong sense, because as well as possessing falsity it also possesses impossibility.

185a31 For none of the others, besides substance, is separate.

Aristotle presents this absurdity as a consequence of the second hypothesis, namely the one that suggests that quantity exists without there being substance; he has not yet set out the consequence of
38,1 the other hypothesis, by which I mean the one that suggests that substance co-exists with quantity. He will speak of that later,[198] though he has already mentioned it in anticipation just before, saying, 'For if there is to be both substance and quality and quantity, and whether these are detached from each other or not, then beings are several'.[199] For in that case, if substance co-exists with quantity, beings are several and not one, and they <sc. Parmenides and his followers> are undermining their own hypotheses.

185a31 For all of them are said in relation to a substrate of substance.

'In relation to a substrate', which Aristotle uses here, is said in place of 'in a substrate'.[200]

185a32 But Melissus says that being is infinite; so being is a quantity; for the infinite is in the realm of quantity.

Since Aristotle has suggested that perhaps they do not mean that all things are substance, but, say, quality or quantity, he is aiming to show by means of these words that it was not arbitrary that he suggested that they meant that being was quality or quantity. For, he is saying, he can show from Melissus' words that Melissus suggested that what he called being was itself quantity; for he said it was infinite, and the infinite belongs to quantity *per se*.[201] Substance or quality or anything else cannot be infinite [*per se*],[202] he says, except *per accidens* in virtue of participating in quantity, just as we say that the white colour is extensive not *per se*, but *per accidens* in virtue of being in a surface that is extensive.[203]

We ought to investigate why it is that he says that the infinite does not exist in substance, except *per accidens* in virtue of participating in quantity, of which the infinite is a *per se* property. For if body is a substance, and its definition is being extended in three dimensions, then body is a magnitude and a quantity; but if that is so, then both the infinite and the finite will belong to it *per se*. And besides, if there are three forms of magnitude – what is extended in one dimension, in two dimensions, and in three dimensions (i.e. line, surface, body) – and body is a substance, then magnitude is a substance. Just as in the case of qualities, there are some which are essential,[204] such as are in the elements, and others which are non-essential[205] and exist *per accidens*, such as in our bodies the whitenesses, blacknesses, instances of heating up that occur from outside, and instances of cooling down, so also there is an essential quantity, which first formats the matter, which is also characteristic of the body. It would require a specialist study to deal thoroughly in more detail with the question, in this connection, whether the property of being extended in three dimensions of a body is really essential, and whether the body is a substance characterised in terms of extension and magnitude, or not.[206] Nevertheless, what we say by way of solving the problem is that, firstly, if body is not a simple thing for Aristotle, but is itself also composed of prime matter and quantity (i.e. what is extended in three dimensions), it is clear that body is not simply quantity.

And besides, 'substance' is not an ambiguous term (it is a genus), but just as 'animal' signifies a common nature that belongs to each of the particular animals (<by this> I mean 'perceptive living thing'), so also 'substance' indicates what is self-supporting. Both bodily substances and non-bodily substances are self-supporting. So if substance is what is self-supporting, but what is self-supporting is not quantity, but the infinite is <a kind> of quantity, then what is self-supporting, that is substance, is not infinite.

Even if we try the argument for bodily substance alone, which acquires the property of being extended in three dimensions as a differentia,[207] even then neither the infinite nor the finite will belong to bodily substance *per se*. For things that are continuous are said to be either finite or infinite in virtue of their size,[208] not in virtue of their being three-dimensional; but quantity in virtue of size is an accidental property of bodies and not constitutive of the essence of bodies.[209] For what makes the plank a plank is not the fact that it is three cubits long, or a digit long, or a thousand cubits. For if 'a cubit's length', for instance, made a defining contribution to there being a substance of plank, there could not be a plank that exceeded, or fell short of, a cubit in extent. And hence also if someone envisaged a plank that went on to infinity, that quantity would not have *per se* being, but would belong to the plank *per accidens*. So that if the finite and infinite belong to quantity in respect of size among things that are continuous, and that is accidental to bodies, then the finite and the infinite belong to bodies *per accidens*.

> **185b3** So if there is both substance and quantity, then being is two, and not one.

Aristotle now introduces the consequence of the remaining hypothesis (the one that says that substance coexists with quantity).[210] In this case, he says, the beings are several and not one.

> **185b4** But if there is only substance, being is not infinite, nor will it have any magnitude. For it will be some quantity.[211]

Having suggested quantity in general, he moves to substance, and suggesting it in a more general sense, simply as substance, he develops the argument in the following way. If they were suggesting that this 'one' was substance, he says, they could no longer suggest that it was finite or infinite; for these are <species> of quantity, but if there is only substance there will be no quantity. So that if Parmenides suggested that being was finite, and Melissus that it was infinite, they cannot have been suggesting that it was substance.

<3.2.1 Exposition and discussion: the structure of Aristotle's argument, 185b5-186a3>

> **185b5** Further, since the one itself is said in several ways, just as being is, we must enquire in what way they are saying that everything is one.

Having declared that he would reply to the followers of Parmenides and Melissus, Aristotle thought he should first tackle the question itself,[212] then using a classification of the one (the one is either *nominally* or *really* one, he says, and if really one, either in respect of the universal or of the particular)[213] he first showed that being cannot be (a) *nominally* one;[214] then he went on to what is (b) *really* one, and (b.1) really one *in respect of the universal*, and he showed that being cannot be one as an attribute either.[215] Next he showed that, from what the followers of Parmenides and Melissus said, they were suggesting that being was a quantity. For Parmenides said that it was finite, and Melissus that it was infinite. So that it was not arbitrary that we suggested that they said that being was quantity.[216] Then finally he showed that they could not be suggesting that it was substance either.[217]

It remains, then, to examine (b 2) <what is one *in respect of*> *the particular*, and this he divides into three. For the particular is (b 2.1) *continuous* or (b 2.2) *indivisible* or (b 2.3) *one in specification*, like things that have several names. Such things are the same as regards the specification of their substance, and differ only in name.

So if being were (b 2.1) one *as something continuous*, since 'continuous' is that whose parts meet at a common boundary (as Aristotle himself defines it in the fifth book of the present treatise),[218] it is clear that what is continuous has more than one part; so that even if it is actually one, nevertheless it is potentially several. But if what is continuous is potentially several, being is not one but several. But those people <sc. the Eleatics> were eliminating plurality entirely, saying that in no way was there plurality in any sense. But if that is the case, they cannot be saying that being is continuous; for what is continuous is potentially several.

If, on the other hand, it is indivisible, like a point or a monad or an 41,1
instant or a change (*kinêma*)[219] (these are the indivisibles among natural things), how will that be either finite, as Parmenides says, or infinite, as Melissus says? What is partless can be a limit, as in the case of an instant or a point, but it can in no way be limited or limitless. For these are certain quantities. So saying that it is one in this way will eliminate the hypotheses of both thinkers at one go.

Further, if the one were partless, it would eliminate both quantity and quality. That quantity will be eliminated is obvious. For quantity has parts, since there are two forms of quantity, continuous or discrete, and both have parts. But they will also eliminate quality, firstly because quality has its being in magnitude, and secondly shape (or form) is a species of quality,[220] and that has parts. So that quality will also be eliminated if being is partless.

But if it is one in specification, like 'people' and 'folk',[221] (these are essentially one, and the difference lies in the name alone, given that the underlying <reality> is the same) then, Aristotle says, 'it turns out that they are uttering Heraclitus' theory'.[222] For Heraclitus said that opposites are the same – white and black, good and bad, sweet and bitter. So now if these people <sc. the Eleatics> were also saying that being is one in this way, in virtue of having one and the same specification,[223] even if it is called by several names, it is plain that they will be reducing opposites to the same thing – hot and cold, dry and wet, good and bad will be the same things, differing in name alone.[224]

Some people, writing in defence of Heraclitus, say that when he said that opposites were the same, he did not mean that good is no different from bad, or white from black, but that (a) they are one in substrate and in kind, and (b) that the opposites in this world are not pure, but somewhat mixed with their opposites. For example it is impossible to find something that is purely good or purely bad; for the ideas themselves of the opposites,[225] being not found in matter, are pure and pristine, just what they are, but when they occur in matter, because matter is receptive of opposites, they are mixed somewhat with each other and no longer remain pure.

But it will be not only the opposites that they reduce to the same thing but also the contradictory; for given that what is not white is something, which on the one hand negates the white, but on the other also signifies something else, if they are saying that all things are one in substance, and differ only in name, then necessarily what is white and what is not white are the same. Hence <both halves of> the contradiction will be true together – the same thing will be white and not white, human and not human, and their theory, he says, will turn out to be not about all things being one, but about there being nothing
42,1 at all. For if the white is the same as the non-white, it will be no more the case that there is white, by reason of the assertion, than that there is not, by reason of the denial. And, further to these things, they will also be reducing substance, quality, quantity, and the ten categories to the same thing; for these differ only in name, not really.

After saying that, Aristotle then adds: it is clear (he says) that we have done well to censure them by taking to absurdity the fact that they reduce opposites to the same thing; this is clear from the fuss made by our predecessors regarding this very puzzle, in the case of what is continuous, lest it should ever be thought that the same thing is both one and several; they took it for granted, on the basis of common sense, that opposites cannot exist together. Zeno the Eleatic, for example, protested against those who had ridiculed his teacher Parmenides' opinion that being is one, and, in defence of his teacher's opinion, attempted to show that there cannot be plurality among the things that exist.[226] For if there is plurality, Zeno says, since the plurality is made up of a number of units, there must be a number of units out of which the plurality is constructed.[227] If, however, we demonstrate that there cannot be a number of units, it is clear that there cannot be a plurality; for the plurality is made of units. But if there cannot be plurality, but necessarily there must be either the one or a plurality, but there cannot be a plurality, we are left with the conclusion that the one exists.

How then did Zeno demonstrate that there cannot be a number of units? Since those who introduced plurality confirmed it on the basis of what is obvious (for there is a horse and a person and each of the particulars, and the collection of these makes up a plurality), Zeno wanted to reject what is obvious in a sophistical way. He said, if the plurality is made up of these things, and a plurality is composed of units, then these are the units. So if we show that these cannot be units, it is clear that there will be no plurality made of them, if the plurality is made of units. He demonstrates this as follows:[228] Socrates, he says, whom you say is a unit contributing to the composition of the plurality, is not just Socrates, but also white, and a philosopher, and pot-bellied, and snub-nosed. So the same person will be both one and several. But it is impossible for the same thing to be both one and several; hence Socrates will not be one. Nor, similarly, will the rest of the things of which you say that the plurality is made up. But if it is impossible for there to be a number of units, it is clear that there will be no plurality either. But if being must be either one or several, but it has been shown that it is not several since there are not a number of units, it must then be one.

Zeno also proves the same thing from what is continuous. For if 43,1
what is continuous is one, since what is continuous is always divisible, it is always possible to cut what has been divided into further portions.[229] If so, then what is continuous is several. So the same thing will be both one and several, which is impossible. So it will not be one. But if none of the continua is one, but it will have to be if the plurality is to be composed of units, since ***.[230] Hence the plurality will not exist either. So we are left with the conclusion that being is one.

In response to this puzzle, Aristotle says, the ancients made a fuss, lest the same thing should ever be both one and several; for they took it for granted from common sense that opposites cannot co-exist at the same time in the same thing. And some, he says, solved the puzzle by eliminating the verb 'to be', such as Lycophron the Sophist;[231] he said that one ought not to say that Socrates *is* white, but that Socrates white; for the word 'is' gives an additional existence to the white. Others,[232] he says, remodelled the language, saying that one ought not to say that Socrates *is* white, but that Socrates whitened, so that the predicate is a verb and not a noun; which is ridiculous, for if 'is' is not supplied in 'Socrates white' (supposing it is not understood from the context) the sentence becomes unintelligible and incomplete. For an assertion is not completed without the inclusion of an indicative verb; in the case of 'Socrates white' it is unclear whether he is or is not <white>. But if 'is' is understood, it will retain the same force as if it had been included from the start in the whole proposition. And saying 'Socrates whitened' or 'Socrates walks', is the same as saying 'Socrates is whitened' or 'Socrates is walking'. For every indicative verb can be analysed into a participle and the verb 'is'.

These are the responses the ancients made to Zeno's puzzles. But Aristotle says that they went through all this and made a fuss lest the same thing should be both one and several, on account of their inexperience in logical method and the significance of words with multiple meanings. What is 'one' is homonymous, he says. For it is <one> either (a) in its subject[233] or (b) in its specification, and again either (i) in potentiality or (ii) in actuality. And likewise for things that are several. So the total number of combinations is six: one and several linked with subject and specification makes four combinations; for they can either be several in their subject and in their specification, or in neither their subject nor in specification, or several in subject but one in specification, or again one in subject but several

44,1 in specification. An example of what is one in its subject and several in specification is what is white and what is hot; for they are one and the same in terms of the subject (for they are both in the same body, e.g. Socrates), but they are different in specification (for the specification of hot is one thing and the specification of white is another). And then again an example of what is one in specification but several in subject is the white in white lead make-up and in snow. The specification (colour that is piercing to the sight) applies to both, but it is several in subject (white lead make-up and snow). A case of what is one in specification and also in subject is clothing and garment, and similarly ground and land.[234] An example of what is several in subject and in specification is wetness and dryness (for they cannot be together in the same subject) and also knowledge and hotness (one is in the soul and the other in the body).

That is how many pairs are produced by linking the one and the several with the subject and the specification; if, however, they are linked with the potential and the actual, they form only two combinations. For they are either potentially one but actually several (like the flames in a number of lamps or the water in a number of jugs; these are potentially one, because they could come together and be made one, but they are actually several because they are divided from each other); or they are actually one but potentially several (like what is continuous). All the other limbs are incoherent; for the same thing cannot be both potentially one and actually so, and similarly with several. So it is far more impossible for it to be simultaneously one and several both in potentiality and in actuality. So Aristotle says that it is an impossible thing for the same thing to be one and several in reality in the same respect; for it is impossible for opposites (or simply contraries) to apply to the same thing at the same time. For example it is impossible for the same thing to be both one and several in its subject, or in its specification, or for the same thing to be potentially both one and several, or actually both one and several; for these are opposites, so they do not coexist. However nothing prevents it being one in one respect and several in another respect, such as one in respect of its subject, but several in respect of its specification (like the sweet and the yellow in honey, or the red and the acidity in wine). And again being actually one but potentially several is nothing absurd (like what is continuous). Or again actually several but potentially one (as in the plurality of quantities of water).[235]

Since, therefore, 'one' is said in several ways, let us solve Zeno's first puzzle[236] by means of the first classification of the one,[237] the one that says it is one either (a) in subject or (b) in specification (the white in Socrates and bald and the rest are one and the same in subject,
45,1 meaning Socrates of course, but a number of things in terms of definitional specification); and <let us solve> his second puzzle about what is continuous,[238] by means of the second <classification of the

one>,[239] by which I mean that of (i) the potentially <one> and (ii) the actually <one>. For what is continuous is one and several, but not in the same respect, but potentially many and actually one. So that opposites will not exist together; for actually one is not the opposite of potentially several, but potentially one of potentially several, and actually one of actually several.

Actually Zeno's argument is self-refuting. As with the universe, he says, given that if it is not several it is one in every way (because of necessity being is either one or several) so also with Socrates and everything else: for if (a) he is bald and snub-nosed and a philosopher and whatever else, and (b) Socrates, being something that is, must necessarily be either one or several, but (c) <given (a)> he is not going to be one, <then> (d) he will of necessity be several. So that in avoiding saying that each of the particulars is one, so as not to say that the one is several by being composed of several units, he will be obliged to say that each of the beings is several. Hence being will be not just several, but several times several.

<3.2.2 Textual analysis and exegesis, 185b5-186a3>

185b7 What is said to be one is (a) what is continuous or (b) what is indivisible or (c) what has the same specification – i.e. the specification of its essence is one.

Something is called individually one, Aristotle says, in three ways: it <is called> one (1) as a continuous thing, as we call the plank one; and (2) as what is indivisible, such as the point; and (3) like things that have several names, like weapons, arms, ammunition.[240]

In what class are we to place Socrates, given that he is one? I say in the class of what is continuous; for the characteristics of Socrates that go to make up him, and the lives themselves,[241] even if not in themselves continuous, nevertheless have their being in something continuous. For this reason we shall put him in the class of the continuous.[242]

185b11 There is a puzzle about part and whole; but perhaps it is not pertinent to our discussion, but is a puzzle in its own right.

Aristotle has shown that if they meant being to be one as a continuous thing, since what is continuous is divisible, being will no longer be properly one, but several. As a result one might have challenged him thus: 'You were wrong to say that what is continuous is several, since 46,1
it is potentially divisible. For what is continuous is the same as the whole, but the whole is the same as its parts; for there is nothing else besides the parts. So what is continuous is also the same as its parts. But if what is continuous is not a number of things but one, since the

whole is as well, it is evident that the parts too will not be a number of things but one; for they are not taken to be actually existent (rather potentially) but in actuality they are one.' So Aristotle says that enquiring into part and whole – whether they are different from each other or the same – presents a puzzle that is not only relevant to the present discussion, but also merits investigation in its own right. For whether someone suggested that they were the same, or different, it appears that some absurdity would follow upon either hypothesis: if the whole is different from its parts, since the whole is nothing other than all the parts, it will turn out that something is other than itself, which is absurd. This is how the syllogism goes:

> The whole is nothing other than all the parts.
> The parts are other than the whole (that is the hypothesis).
> Hence the whole is other than the whole.

Hence the whole is other than itself. So that is what would follow if the hypothesis were that it was other than its parts.

But if the whole is the same as the entirety of its parts, it would also be the same as each one of them; for if it were not the same as each one of them, nor would it be the same as all of them together. But if it is the same as each one of them, then the parts will also be the same as each other (which is absurd), for things that are the same as the same thing are the same as each other. So that as far as these puzzles go, the whole will seem to be neither the same as, nor other than, the parts.

We must understand, therefore, that the whole is in one way the same as its parts and in another way distinct. For it is distinct from its parts as <something> supervening on the assembly of all the parts, and <something that> is the completion of that assembly; but then again the whole is the same as its parts in that nothing supervenes from outside the range of the parts to contribute to the existence of the whole. I will give an example: if you tune each string of a lyre individually to the note that happens to belong to the Lydian chord,[243] and they are all separate and not conjoined into one, clearly each one produces its own note by itself; but the chord made with the combination of all the strings is produced out of the notes of all of them, and is clearly distinct from the note made by all of them before they came together. The concurrence of them all has produced a certain form, which was not there among all the particular notes before they came together. Thus the entirety of the chord is distinct from all the particular notes, even if they sounded at the same time apart from each other, but is the same as them in that nothing else
47,1 besides the particular soundings contributes to producing the form of the chord.

The same applies to a human person. If you think of the parts of the person torn apart, but each one having its proper function (e.g. the eye having the function of an eye, and the hand and the rest likewise), while these are all separated, the form that is produced from the concurrence of all of them is something different. But it would also be called the same thing, since nothing else besides the several parts and their proper functions goes into producing the entirety of the form; for the common function of the organism is a product of the functions of the parts.

> **185b11** But perhaps it is not pertinent to our discussion, but is a puzzle in its own right.

Why does Aristotle think that the puzzle about the whole is not pertinent to the discussion before us? And if it is of no relevance to the <discussion> before us, why does he mention it? It will seem irrelevant to the <discussion> before us, because when the issue is whether being is one or not, and what has just been said is 'if it were one as what is continuous, since what is continuous is divisible *ad infinitum*, it will be several', the puzzle about the whole and the parts will not help either to confirm or refute this for us. For if it be the same <as its parts>, being will not on that account be one (for even so it can be divided *ad infinitum*), or if it be not the same but different, then for this very reason the beings are already several, unless we take it as being divisible. In general the puzzle as to whether it is the same as its parts or different does not seem to belong to the present discussion. Why Aristotle mentions it and how it has bearing on the subject before us we said at the beginning.[244] And in general, since he had mentioned the division of what is continuous into parts, this sort of issue <is> not unconnected with the <discussion> before us;[245] for someone who says that something is divided into parts must have a notion of the nature of the whole and of the parts.

> **185b12** Whether the whole and the parts are one or more than one, [and how they are one or more than one, and if more than one, how more than one, and about the parts of things that are not continuous].

Aristotle has put 'one' instead of 'the same', and 'more than one' instead of 'different'; and 'how they are one or more than one' instead of 'how they are the same and how different'. He says 'and about the parts of things that are not continuous' either about parts in juxtaposition or about discrete parts. Parts in juxtaposition are as in the case of a house; for here too we must ask whether the whole is the same as its parts or different; if it is the same it is also the same as each part, but if so they will be the same as each other, so that the roof is the same as the foundation, and the foundation is the same as the 48,1 walls, which is absurd. By discrete parts I mean like foot, hand, and the rest. For if the whole is the same as each of the non-uniform parts, they will also be the same as each other. Hence the hand will be the same as the foot.

> **185b15** And if either <is> one with the whole as an indivisible thing, that they <are> also themselves with themselves.[246]

If either one of the parts, Aristotle is saying, is the same as the whole, because it is indivisible, it is evident that they will also be the same as themselves. The phrase 'as an indivisible thing' is an explanatory clause, explaining why the part is the same as the whole. 'Either' is in place of 'each'; that is, because the parts of the whole are indivisible, each of the parts is the same as the whole ('one' again stands in for 'the same'); if this is so, the parts are also 'themselves with themselves', that is they will be the same as each other; so the head will be the same as the hand. So that is the sense of the passage if we understand 'either' to mean 'each'.

But if 'either' is taken in its usual way to apply to two things, it would mean the whole and the part. The idea is as follows: since the whole is indivisible, and in that whole is seen also the part, on this account the part is the same as the whole; if this is so, then also the parts will be the same as themselves (that is, each other). But why are the parts also the same as each other if the part is the same as the whole? Because if each part of Socrates is the same as the whole Socrates, and things that are the same as the same thing are the same as each other, then the parts of Socrates will be the same as each other.

> **185b16** But if it is as something indivisible, there is no quantity or quality.

Having refuted the suggestion that being is one in the manner of what is continuous, Aristotle goes on to the suggestion <to the effect that being is (b 2.2) one> as a partless thing,[247] and says (as we have already said)[248] that they will be abolishing both qualities and quantities, and that they will no longer succeed in speaking of it as infinite or finite.

49,1

> **185b19** But if <one> in specification, like clothing and garment ...[249]

Aristotle has moved on to the third suggestion,[250] the one that says that <being> is one in its definitional specification but not in its names, like things that have several names.

185b21 For what it is to be good will be the same as what it is to be bad

in that they will reduce opposites to the same thing, like Heraclitus.[251]

185b22 And to be neither good nor bad[252]

in that they will also reduce the contradictory to the same thing, so that <both halves of> the contradiction will be true together.

185b23 And person and horse

in that they also reduce things that are simply different.

185b23 And the argument will not be about things being one, but about nothing.

For if <both halves of> the contradiction are true together, someone who says that being is one will be saying that being is nothing. So that the argument starting from these <claims> risks proving that being is no more *one* than it is *nothing*.

185b25 And what it is to be of a certain sort and what it is to be of a certain quantity will be the same.

For if things differ only in name, quantity and quality and the rest of the categories will be the same.

185b25 And the more recent among the ancients made a fuss.

This does not relate to what follows but to what came before, when Aristotle said 'For what it is to be good will be the same as what it is to be bad'.[253] So absurd is it for opposites to be the same, he is saying, that our predecessors were alarmed lest the same thing should seem to be one and several at the same time, such as Socrates being both white and a philosopher and an Athenian.

185b28 Others remodelled the language

like Menedemus of Eretria.[254] Aristotle says this without naming anyone, and some,[255] who are entirely unfamiliar with Platonic usage, have supposed that he was hinting at Plato. For, first, the word 'is' (*esti*) is always being additionally applied to premisses: 'there is something, that which is just' and 'there is something, that which is beautiful',[256] and, second, in the *Sophist* he mentions this very theory, and rebukes those who think that the same thing is one and several

50,1 because more than one thing is predicated of it.[257] He says they are afflicted this way due to their lack of acquired sense, they being unable to distinguish the different ways in which being one and being many apply to the same thing.

185b31 As though 'one' or 'being' were said in only one way.

Aristotle says this in the middle by way of mockery of them. 'As though "one" was said in one way and not several ways', he says, like 'being' too; for the phrase 'or "being" ' stands for 'like "being" too'.

185b32 Things are several either in specification ...

What is musical and what is white are different in their definitional specification, but the same in their subject; for they are in one and the same person.

185b33 So the one is several

What is one in subject is several in specification, and this is not absurd, nor impossible.

185b34 Or by division, like the whole and the parts. [But they were at a loss at this point, and admitted that the one was several – as if it was not acceptable for the same thing to be both one and several, provided they are not the contraries.]

Aristotle says 'by division' instead of 'potentially'. What is continuous is actually one thing as a whole, but potentially several, and that is not absurd. But the ancients, he says, 'were at a loss at this point', that is concerning what is continuous, and, as if overpowered by the difficulty, rushed into the arms of absurdity, and 'agreed that the same thing is both one and several' – as though it were 'not acceptable for the same thing to be both one and several', he says, providing they are not the contraries. For it is impossible for the same thing to be potentially both one and several, or for the same thing to be actually both one and several (for these are contraries, and contraries cannot co-exist), but for the same thing to be potentially several and actually one is no impossibility. That is what the continuous thing is like.

<4 On Melissus: Book 1 Chapter 3, 186a4-22>

<4.1 Exposition and discussion: an analysis of Melissus' reasoning, 186a4-22>

186a4 When one approaches it in this manner it looks impossible that things should be one, and there is no difficulty in releasing oneself from the <premisses> from which they <sc. the monists> derive their conclusions.

Having earlier addressed himself to the doctrine itself,[258] Aristotle now turns to the actual arguments of the members of the school of Parmenides and Melissus, the arguments by which they supported the claim that the universe is one and motionless. First he deals with Melissus' argument. We shall need to set out the whole of Melissus' argument first, and then examine each of the premisses in the manner in which they were originally set up by Melissus, and finally say how Aristotle knocks them down.[259]

Melissus' argument is as follows – or rather the syllogism is not Melissus', but he went so far as to posit that being (*to on*) is without beginning, being is one, being is infinite, being is motionless. These four theses were posited by Melissus, and some have thought that he deduced such <theses> on the basis of a syllogism, which Aristotle reports.[260] Whether Melissus also provided the syllogism, or whether others thought that he deduced these <theses> on the basis of some such syllogism, will make no difference to us. In any case the syllogism from which the said <theses> are deduced goes as follows:

Being did not come into being.
What does not come into being has no beginning.
What has no beginning has no boundary.
What has no boundary is infinite.
So being is infinite.

So in this way he proved that it is infinite; and hence that it is also one and motionless. For if it is infinite, it is clear that it would also be one; for what is infinite takes up all the room, and if so it will not allow there to be anything else, since if it did it would not be infinite. But because it is one and infinite, by reason of both these it will also be motionless. For if it has taken up all the room, it will not have anywhere to move. And again if it is one, it must also be motionless; for if it moved there must be either a place or a time in which it moves, or the form of the motion with which it moves; hence there will now not be one but several things. By these <arguments> Melissus establishes that being is one and infinite and motionless.

We ought to investigate each of the premisses, to see where he got it from.

(1) He shows that being did not come into being as follows: if being came into being, it came either from being or from not-being. If it came from being, being will be there before it comes into being (for what is something comes from what is something, but being, *tout court*, could not come from being; it would be coming to be from itself, so that it was before it came into being, which is absurd). If it came into being from not-being, it must be from what is not in any manner or form; for if being *tout court* comes into being, it would come into being from not-being *tout court*. But there is a common consensus among all the natural philosophers, that nothing comes to be from what in no way and in no sense is.[261] For something must remain, in the process of becoming, that does the changing, and what becomes must have the potential to become that which it will become; for the child must have the potential to become literate. So therefore one who is to become a person must have some suitability for that, but the thing that has the suitability, and indeed the suitability itself, is something that is; so everything that comes to be comes from something that is. So therefore if it is necessary that if being comes to be it must come either from being or from not-being, but absurdity follows in either case, clearly then, being did not come into being.

(2) That what does not come into being does not have a beginning, Melissus demonstrates from the contrary by using the method of conversion.[262] He takes the premiss that everything that comes to be has a beginning. If then what comes to be has a beginning, what does not come to be does not have a beginning.

(3) That what has a beginning has a boundary as well he takes from common consensus again; for the beginning itself is a boundary.

This is the whole of the proof for the Melissian argument.

Aristotle makes a number of charges against Melissus, and firstly against the conversion, on the grounds that Melissus has done it incorrectly, from the antecedent. For the rule of conversion by negation[263] ought to be done from the consequent, for then it becomes true, when we take the contrary of the consequent and prove the contrary of the antecedent.[264] But if the conversion occurs from the antecedent, it does not invariably come out true, unless the terms in the conversion are coextensive. If human, then invariably also animal;[265] but it is not the case that if something is not human then for that reason it will invariably be not animal either. For the horse is not human, but nevertheless animal. But if something is not animal, then it is not human either. So converting from the consequent is always true, but the conversion from the antecedent is false.

If, then, we convert from the consequent, as is the rule for conversion by negation, which always gives what is necessary and true, we form a syllogism in the second figure from two negative premisses: 'For if what comes to be has a beginning, what does not have a beginning did not come to be. Being therefore did not come to be.

What does not have a beginning did not come to be.' So that the sequence does not form a syllogism.[266]

Secondly, even if we allow Melissus the conversion, his argument will be refuted as follows: there are two senses of the word 'origin' (*arkhê*); firstly the temporal beginning – the very moment in which something comes into being is called origin (*arkhê*) – and secondly the real principle – such as matter or form – is called origin (*arkhê*).[267] That is, in an animal, what nature makes first, say the blastocyst,[268] or alternatively the heart or the brain is an *arkhê*. 53,1 So what sort of an *arkhê* did Melissus have in mind? If it was temporal let us see what follows from the argument. Being did not come into being. What does not come into being does not have a temporal beginning; what does not have a temporal beginning does not have a temporal boundary (for clearly we must take the word 'boundary' in an analogous sense to the sense of '*arkhê*'); what does not have a temporal boundary is endless, so being is endless. But then it is also without beginning in time since it did not come to be, and what is without beginning or end is everlasting, so being is everlasting. So the consequence of the conversion, if we take *arkhê* as temporal, is that being is not infinite but everlasting.

If, on the other hand, we take *arkhê* as the real principle, it loses its capacity for conversion; for it is not the case that if what came into being has a real principle, then what did not come into being will not have a real principle. For what is there to prevent it from having as its real principle its matter and its form, if being is still a body, even if it was without origin? Spheres would have their centre as their real principle and their surface as their boundary, even if they were without origin and someone were to concede this.

Thirdly, Aristotle says, what is there to prevent it, infinite as it is, from moving by revolving round itself and not taking up any further room, like the water in the basin and like all the things that move in a circle, for example the heavenly spheres? And fourthly, he says, even supposing the infinite cannot change in respect of place, what is there to prevent it from changing in respect of alteration, on your assumptions? Yet Aristotle himself shows that the infinite, supposing it were to exist at all, would also have infinite power in every way and in every respect, and for this reason it would be changeless in respect of every kind of change; for if it changed it would have to change under the influence of a thing that produces change; but what is affected is affected by some greater force (for something would not be affected by a lesser one or by an equal one), but there is nothing greater than the infinite. And he shows that there will be nothing to act upon the infinite either, if nothing else can exist along with it given that the infinite has taken up everything. For it would not be infinite in one dimension but not in another (that is the stuff of myth, and arbitrary), but it is clear that, if a body could be infinite at all, it would be infinite in the three dimensions. So Aristotle, as I said, demonstrates that the infinite is also changeless in every way, because it is also of infinite power; but Melissus said that the infinite did not move for one reason alone, that it did not have anywhere to move because it took up all the room, and Aristotle therefore, in responding to that suggestion, says that nothing prevents it from locomotion by rotating, and from alteration. If it was for that reason alone that Melissus said that being was motionless, in fact nothing prevented it from moving in a circle and altering; but if it was not just for that reason, but because he also said that it was one, he would not be refuted as far as what we can get out of these arguments. For what is actually one and only one would be unchangeable as regards every kind of change. For if it changes there will be something in respect of which the change is, and thus it will not be one but more than one again.

These are Aristotle's charges against Melissus' argument. He does not argue against the claim that being did not come into being; for he too thinks that is so, since he believes the universe is without origin. However I think it is reasonable not to leave unexamined Melissus' attempted proof that being could not have come into being.[269] I think it is obvious to all that being *tout court* would not come to be from being; for it would be there before it came into being. But why is it also impossible for it to come from not-being? Just because it seemed so to most of the natural philosophers, it does not follow that it is necessarily true. For many of their other opinions were not true, and a lot of those Aristotle too refuted.

If we say that before it came into being it was capable of coming into being, we do not thereby suggest that its origin was from some pre-existent entity serving as material cause. What prevents us from saying that before it came into being it was capable of coming into being because its efficient cause pre-existed, and not because any matter was supplied in advance and changed into what came into being? It is not appropriate to apply the word 'capable' to the material cause, unless it is shown to be without origin in its own right; for how could one say that matter is capable of being a 'this', unless it were first shown that this thing itself subsisted? Just as the form would not be said to be capable of anything before it subsisted. So that if someone said that it originates from an entity rather than from an efficient cause, we should agree that every thing that comes into being originates from an entity (for it is impossible for anything to have an origin without a cause) but if they mean from an entity as material cause supplied in advance, it is no longer necessary, as a consequence of saying 'being was capable of coming into being before it came into being', that it originated from an entity; rather from not-being, unless it has first been shown that matter itself is without origin and indestructible, by its very specification.

So far, then, the proof just cited does not entail that things that come into being necessarily originate from being in respect of the material cause, but as far as that <cause> is concerned coming to be from not-being is also allowed, matter being called into existence 55,1 along with the forms by the creative cause. For just as it <sc. the creative cause> introduces the forms which were not there before, what prevents it from introducing matter too which was not there, so that it may be creator (*dêmiourgos*) *in toto*? But if it <sc. the creator> also introduces matter, that too must be either from being or from not-being, and this goes on *ad infinitum*.

But perhaps in the case of being *tout court* there is no room at all for inquiring whether it came into being or not, if being is not a kind among the ten categories, but an ambiguous term divided into different meanings. The genera, and universals in general, indicate a certain nature; for the nature of animal is a different thing from human and from each of the others; hence even if it is invariably seen in particular animals, yet it has a nature according to its own specification, whether it came into being or not. But terms that are ambiguous do not indicate a substance or any common entity at all, so how could they be said to come into being or not? The word 'mouse', say, is not indicative of any common nature.[270] So given that there is no common underlying nature for ambiguous terms, how could one say either that mouse came into being or that it did not? In the same way if being is not a genus, but only an ambiguous term, clearly no common nature is signified by 'being'. For the connection between the ten categories is only verbal; there is no such thing as a being *tout court*, but it is either substance or quantity or one of the others. So it does not make sense to inquire about it, either that it did come into being or that it did not; but we do inquire whether body came into being or not, whether the world came into being or not, and if it did, whether it came from beings or non-beings. And it is not impossible to say, about these, both (a) that they came from beings, if it were shown that the matter, in which what came to be had its subsistence, was supplied in advance, and (b) that they came from non-beings, if matter itself was also brought into existence at the same time as the form. Such are the responses to be advanced on our part,[271] for the present, against the claim that being did not come into being; given that theories of this nature have been adequately explored by us elsewhere.[272]

But since Aristotle too agrees that these men are not natural philosophers, and indeed the men's own writings show that – for Parmenides says in the *Towards Truth*[273] that being is one, and in the *Towards Opinion* he says there are two principles of things, fire and water,[274] incorporating the intermediate things as well via the extremes, so that in these latter verses he is actually talking about physical things (which they also call 'opinables' (*doxasta*))[275] while in the *Towards Truth* he is discussing intelligible things – for this reason it is worth investigating, in relation to what the man himself had in mind, the implications of his premisses.[276]

56,1 One should be aware that they[277] called everything intelligible 'being', and what is perceptible, 'becoming'. Indeed, that is Plato's terminology too, inherited from him;[278] Plato says 'What is it that is always in being, and has no coming-to-be?' about the intelligible; and 'What is it that is always coming to be, but never in being?' about the perceptible.[279] 'The one is grasped by intelligence with reason, the other opined by opinion with irrational perception.'[280] They found that unification predominates in the intelligible world – for there is unification and diversification in all things, but what is stronger and closer to the principle of all things is always more dominated by unification. Indeed there is both unification and diversification in us and in the irrational animals, but unification predominates more in us than it does in them; for the human being is a tame and social creature. But diversification predominates more in them. And then again unification and diversification are seen both in us and in the heavenly bodies, but unification predominates more in them than in us. In the same way, unification has predominance among the intelligibles, more than in anything else.

Since, therefore, they found that the intelligibles were dominated by unification, and, as it were, that throughout everything the one common principle of all things was seen through union with it, they said that everything intelligible was one and without beginning, not distinguishing them <sc. intelligibles> from their own principle, nor recognising that the principle is one thing and what derives from it another, because throughout everything it was seen to be one and without beginning. Consequently they said that it did not have a boundary, because it was infinitely powerful and extended to everything and was not bounded by anything. So that if they called everything intelligible 'being' and not becoming, and said that everything was without beginning for the reason just given, then these two will be coextensive – 'not becoming' and 'without beginning' – for both apply to the same things. Similarly if they said that every perceptible came into being, and every one of these has a principle (*arkhê*), whether chronological or real, then these terms will also be coextensive – 'come into being' and 'having an *arkhê*'.[281] So that if something is come into being, it will also have an *arkhê*, and if something has an *arkhê*, it also came into being. And again if something did not come into being, it is something that does not have an *arkhê*, and again if something does not have an *arkhê* it did not come into being. So that
57,1 since these are coextensive, and when things are coextensive it makes no difference whether the conversion is done from the antecedent or the consequent, it was not unreasonable that Melissus, considering the conversion to be applied to things that are coextensive, reckoned it to be indifferent that it was converted from the antecedent.[282] For

example in the case of risible and human it makes no difference to say 'If human then risible; if not risible, then not human either' rather than 'but if not human, then not risible either'.[283]

He also meant it was motionless in this sense, in that the being and activity that those <intelligibles> had was stable and entirely changeless. Aristotle too calls all the intelligibles motionless in this sense, not only divine things but also souls, in the treatise where he demonstrates that the primary movers must themselves be unmoved.[284] It is right that Aristotle, writing as a physicist, takes the arguments in a more physical sense, and that Melissus used such arguments in a more theological way.

<4.2 Textual analysis and exegesis, 186a4-22>

> **186a4** When one approaches it in this manner it looks impossible that things should be one, and there is no difficulty in releasing oneself from the <premisses> from which they <sc. the monists> derive their conclusions.

Having set out to refute the view of the school of Parmenides and Melissus, Aristotle first explored the question in its own right in a more general way, and is now going to refute Melissus' argument. For this reason he says that someone who proceeds in the manner spoken of will find this sort of view – about all things being one – is impossible, and that the arguments by which these people tried to prove it are not difficult to refute. The expert must not simply demonstrate things in themselves, how they stand in nature, but also resolve the difficulties that are brought against the correct view of them. For whichever of these is lacking, the other is crippled; if you refute the arguments of the opponents it remains unclear how things stand in nature – for one can argue in support of something true but not from premisses that are necessary (e.g. supporting the idea that the soul is immortal from the fact that one respects the graves of one's forebears); someone who refutes those arguments has not yet disproved the doctrine. What's the evidence that there are not other arguments that support it from necessity? Or if we demonstrate things in themselves, how they stand in nature, but we do not resolve the puzzles (*aporiai*), the puzzles will remain to trouble us. For this reason Aristotle, having shown that being cannot be one, now also sets out for refutation the arguments that support it. 58,1

> **186a6** Both thinkers, Melissus and Parmenides, reason in an eristic manner,

he says, adopting false premisses and also reasoning invalidly. He charges them with invalidity, on the grounds that both of them use conversion incorrectly, doing it from the antecedent. For Parmenides also uses conversion in the same way. 'What is apart from being', he says, 'is not-being; not-being is nothing <i.e. not one thing>; so being is one'.[285] The inference drawn from the premisses is 'So what is apart from being is nothing'; but he has converted this conclusion, but converted it from the antecedent when it should be converted from the consequent: 'For if that which is apart from being is nothing', he says, 'then being is one'. For being is the contrary of non-being, and one is the contrary of nothing. The correct thing would be to convert from the consequent and say 'So the one is <a> being'.

186a8 Thus their arguments are invalid,

in that they employ conversion incorrectly. 'But they adopt false premisses'; for now, Melissus, in that he thinks that what does not come into being does not have an *arkhê*. For even if this premiss was obtained by means of the conversion, it is possible to derive a true conclusion from invalid reasoning (e.g. when I say 'A person is not a stone; a stone is not a horse; therefore a person is not a horse'; here the conclusion is true, but it is derived invalidly from two negative premisses). So since it is not invariably the case that what follows from invalid arguments is false, for this reason Aristotle not only accuses him of obtaining the premisses invalidly, but also that in addition to the invalidity, he has also got *false* premisses.[286]

Parmenides is also in the same position. He deduces something false, that being is one, by means of the invalid conversion. But Aristotle says that Melissus' argument is more shoddy, because Parmenides gave two arguments, and in one of them, to which we have just been attending, he takes true premisses (what is apart from being is not-being; not-being is nothing), but converts the conclusion incorrectly. Melissus, on the other hand, uses conversion by negation incorrectly right at the very beginning, and as a consequence of this adopts false premisses. So it is shoddy because right from the very acceptance of the premisses it is offensive to the listeners, and it says nothing that is clever or capable of posing a difficulty.

Parmenides, however, has a sharper approach in the second argument as well. This is how the argument goes: 'If "being" signifies one thing, and the contradictory cannot be true at the same time, then being is one.' He has not structured the premisses so that they connect at a common term, nor has he derived the conclusion from the 59,1 terms that are listed, and for this reason, since its refutation is not apparent it would most likely pose a difficulty for the listeners, as to what truth or falsity there was in what was said. For this reason Aristotle said that Melissus' argument was more shoddy, because it offers no difficulty, but is immediately and easily shown up by the listeners. For we do in fact think such arguments are shoddy and a right pain, arguments that, as well as being false, are neither sharp nor difficult to refute.

186a9 But once one absurdity is granted, the rest follows.

For since Melissus accepts one absurdity, the invalid conversion, right at the beginning, and from this the false premiss that says 'what does not come into being does not have an *arkhê*' – real *arkhê*, evidently – in this way he deduces the ensuing absurdities from that beginning, that being is one and infinite and motionless. But this, he says, is nothing difficult, that someone who asks for one false thing, should derive a myriad false things from it. If it were granted that human beings are birds, it will follow that they also have wings and fly and are irrational and lay eggs and a myriad other things.

186a13 And then this is also absurd, to think that there is an *arkhê* of everything, in the sense of a real principle and not a chronological beginning.

It is evident that Melissus took the *arkhê* as real, but the word *arkhê* is ambiguous. Aristotle teaches us, in the place where he treats the topic of homonymous terms, that when an ambiguous term is used in a question, one must distinguish the things signified by the ambiguous term, and determine which of them the argument is about.[287] For this reason he criticises Melissus because when he was discussing an ambiguous term, *arkhê*, he did not distinguish the things it signified, but discussed it as if it signified one thing, the real principle.

186a14 And of coming to be, not *tout court*, but of alteration as well, as though change did not occur in a rush.

Melissus was suggesting that everything that comes into being *ipso facto* also has a real principle, and that the real principle is the partile principle.[288] For it could not be simply the elemental principle, for that way he has not inferred what he wanted. For even if some things had not come into being, but were nevertheless bodies, they must surely invariably have the elemental principle; for every body consists of matter and form at their simplest. Furthermore it is not the 60,1
case that if something lacks the elemental principle it is necessarily infinite in size. On the contrary, if something is not composed of elements it will not have any size at all. So that, rather by far, it will be neither finite nor infinite. All bodiless things are like that, for example soul and mind, and, equally, the point is like that. So he took the principle to be the partile principle, for instance that this part, say, of every thing that comes into being is the first that comes into being.

Aristotle criticises him, on the grounds that he wrongly thinks that not only (a) the principle of absolute coming into being is of that sort, but also (b) the principle of absolutely every coming into being is of that sort. By 'absolute coming into being' he clearly means the coming into being of substance. For there is alteration, he says, that does not occur part by part, but the thing alters suddenly, as in the curdling of milk; it is not that one part of it changes first and another part later, but the curdling occurs through and through without any rank order. Similarly in the case of suntan – the parts which are exposed to the sun in the first place darken at the same time altogether without any rank order, not this part first and another one second. So he was wrong in supposing that there is the same principle for every thing that comes into being. For there is a principle in a different way even for the case of sudden alteration where the form does not arrive all at once suddenly, as it does for complete curdling, but where some traces of the form arrive in advance – I mean gradual curdling.

> **186a16** And then, why is it motionless if it is one? Just as the part that is inside, this water, moves within itself, why not the whole thing too?

What prevents being from moving, Aristotle is saying, even if it is one? Just as the part of the water, which exists within the whole, itself moves, even though the whole remains unmoved – as for example when we dabble our hands in the sea, and swirl the water round our hand, that's what the whirlpools that occur in the water are like too – what stops being, too, infinite and one though it is, from being unmoved as a whole, he says, but moving as to its parts? Just as the
61,1 universe, too, is motionless as a whole but moving as to its parts.

This is the meaning if we read the phrase '*hôsper gar kai to meros enon*'[289] with the word '*enon*' being taken as one word, equivalent to 'being within'. But if, on the other hand, we read it with a space[290] the phrase is equivalent to: 'Just as the part of the water, I mean this water, which is itself one as well, moves within itself …'. For the part that is separated from the whole, e.g. the water in the basin, is also itself one thing. So just as this is also one thing, and nothing prevents it from moving within itself, why is it not the case that being, too, even if it is one, moves within itself?

And then, Aristotle says, let us grant that it does not change place; what prevents it from changing by alteration? But we maintain that the unity thesis eliminates change altogether.[291]

> **186a19** But nor can it be one in form, except as regards what it comes from.

As though defending Melissus' opinion, Aristotle says that if one is to say that being is, in fact, one, it will be not in form but in substrate that being is one in this way, just as some of the natural philosophers said, for example those who spoke of atoms (the atoms are one as to their kind, but they differ in shape) and those who suggested that the element was one, and generated other things by condensation and rarefaction of this element. But even if things are one in their substrate, nevertheless they differ in their forms. So that things will not be one on account of their substrate any more than they are several on account of their forms, particularly as things are characterised not by their matter but by their form.

<5 Parmenides and Zeno, Book 1 Chapter 3, 186a22-187a10>

<5.1.1 Exposition and discussion: the structure of Aristotle's argument, 186a22-186b14>

186a22 And the same way of arguing applies to Parmenides too.

Aristotle has attacked the question in general, on the grounds that it is impossible for being to be one, and then examined Melissus' argument in support of the view that being is one, and refuted it; next in the sequence he now examines Parmenides' arguments and refutes them. Just as the views of Melissus and Parmenides have some features in common (saying that being is one and motionless) and some specific to each (one says being is infinite, the other finite) so also the refutations used against them are some of them common <to both> and some specific to each. The common ones are the ones that 62,1 prove that being is not one and motionless, and the specific ones are the ones that refute Melissus – proving that it is not infinite – and Parmenides – proving that it is not finite either. Aristotle is going to show that this being, which they also take to be one, does not admit of being infinite, nor of being finite either.

Parmenides has two arguments, however, one of which is reported in this passage (and is also mentioned by Theophrastus[292] and Plato[293]); the other has been reported earlier.[294] The argument[295] goes like this. 'What is apart from being', he says, 'is not-being, not-being is nothing; so one is being, or alternatively, being is one'.[296] That is how the argument goes, but it fails both in respect of its premisses and in respect of the conversion; for the conclusion is converted incorrectly from the antecedent. The conclusion is 'So what is apart from being is nothing'.[297] While conversion by negation is universally true whenever we take the contrary of the consequent and infer the contrary of the antecedent, Parmenides has carried out the conversion incorrectly, taking the contrary of the antecedent and inferring the contrary of the consequent. In the conclusion the antecedent is 'What is apart from being', the consequent is the expression 'nothing'; 'being' is the contrary of 'what is apart from being, and 'one' is the contrary of 'nothing'. If he had done the conversion correctly , he would have inferred from the consequent 'the one then is <a> being', and that would have been true. Such is the error in the conversion of the conclusion.[298]

Aristotle also objects to the premisses, and firstly to the minor premiss that says that 'What is apart from being is not-being'; for since it is evident, he says, that the things that have being are several (let this be taken on the basis of the evidence) the argument ought not to be developed in the singular, nor should it accept an ambiguous word (being) in the premiss, but it should say 'What there are apart from the things that have being are things that do not have being'. But if he had said that, he would *ipso facto* have admitted that the things that have being are several.

However, even if we grant him that 'being' signifies one thing and that the minor premiss is true, the major premiss is false. Let it be true that what is apart from being is not-being, 'being' signifying one thing, such as substance; yet it is after all no longer true that not-being is nothing; for attributes are not-beings, in that they are not substances, but they are not nothing.

Plato also attacked this premiss in his dialogue the *Sophist*.[299] Having first sung the praises of Parmenides, and called him 'father', 63,1 he says that even if it is true that what is apart from being is not-being – and we know that he called the intelligible 'being'; so if what is apart from the intelligible is something, that will be not-being – yet nevertheless it is not true that not-being is nothing. For all perceptibles are non-beings (because only intelligibles are beings) but they are not nothing. For even the perceptibles are something. So that even if we suggest that the minor premiss tells the truth, the major premiss is false. But nothing follows from one premiss.

<5.1.1.1 The meaning of 'one'>

But besides, in what way is he saying that being is one? For it will be either nominally one or really one. And if it is really one, it will be one either generically or specifically or numerically.[300] And if it is numerically one, it will be either as what is continuous or as one in specification; for at this point he leaves out one as indivisible, because there is no room for that suggestion in the case of these men, if one of them says that being is infinite and the other says that it is finite, and the infinite and the finite belong in the field of magnitude, but the indivisible is neither finite nor infinite.[301] If, therefore, being is one only nominally, and there are different real things, then the things that have being are *ipso facto* several.

But if it is really one, whether generically or specifically, again the things that have being are several; for the generic kind encompasses several things that differ in species, and the species encompasses several things that differ individually. But if being were numerically one, if it is the same in specification (like something with several names, I mean) they will be uttering Heraclitus' theory (this Aristotle has already said above)[302] in such a way that opposites will be the same, and contradiction, and being will no more be than not be. But if being is one as what is continuous this will have to be either just substance, or just attribute,[303] or a composite of substance and attribute, Aristotle says. If it is a composite of substance and attribute, obviously the things that have being are several. For since substance and attributes coexist with each other clearly they cannot be one and the same thing. For this, he says, is indeed the very cause of Parmenides' error: he did not recognise that the recipient of the attributes is one thing and the attributes themselves another, not just in relation to the substrate but also in relation to each other. For all these things, even if they go to make up one thing, nevertheless the concept of the essence of each is different. You will give the definition of the substrate (by which I mean the substance) in a different way from the way you would give it for the quantity and the quality and the rest. So if they were to suggest that what is continuous was one in this way, as a composite of substance and attributes, obviously again being would be several and not one.

But if they were suggesting that being is substance only – which it is plausible to suppose <that they were suggesting>, for they certainly did not say that this being was an attribute ***[304] then having mentioned substance, and observing that it is not plausible to suggest that they were saying that being is an attribute, Aristotle first questions this latter part of the classification,[305] and then goes back to the suggestion about substance.[306] 64,1

<5.1.1.2 The one as attribute>

For, Aristotle says, if we were to say that being is an attribute, given that an attribute is an attribute of something (of substance in fact), there must exist this something of which it is an attribute; so that will be substance. But since attribute alone is being, it is clear that substance will be not-being – not some of it being and some of it not-being (for in that way there would be a plurality of beings), but wholly non-being. But on the other hand it was supposed to exist as well, so that being (i.e. attribute) could indeed be an attribute of it; hence some non-being will exist, which is absurd. For despite (a) not-being – because attribute alone is being, still (b) there must exist the substance, of which being is said to be the attribute; so there will be something both being and not-being, namely substance.

In another way too it will be both not-being (given this hypothesis) and also being, given that it participates in being (i.e. attribute) and things that participate in something also acquire the name derivatively from what they participate in. For what participates in the white is said to be white, and what participates in the musical is said to be musical: '*mousikê*' is the word both for the expertise and for a musical woman. Hence if not-being, that is substance, participates in attribute, that is being, then substance too would be being. Hence the same thing will be both being and not-being.

Furthermore since attribute is being, but this being is an attribute of something else, but that other is not-being, then being will be the same as not-being; for it is in not-being, and what is in not-being will be not-being.[307] So it is impossible that being is an attribute.

Having demonstrated this, he returns to the suggestion at the beginning[308] that perhaps they are suggesting that being is substance alone 'such as a single man' or simply a body. But if this is so, he says, given that attributes belong to substances, and as a result it is correct to say that Socrates is white, or the man, or the body, again the same thing will be both being and not-being. For if it is correct to say that the substance is white, but the white is not-being, then the substance will be not-being. But it is also being. So being and not-being will be the same. And again if the white is not-being and this is in the substance, but the substance is being, then in being there will be things-that-are-not. But what is in being is being. So not-being will be being. So they cannot be suggesting that being is substance alone.

<5.1.1.3 Parmenides' text>

Perhaps, therefore, they are not suggesting that there are any attributes at all, but only substance alone by itself, and thus none of the aforementioned absurdities will follow. Aristotle tries out this suggestion next; but it is our duty to listen sympathetically to Parmenides' own words and to grasp Parmenides' own meaning from his original text. So what does he say about being in his verses?

> Whole, sole in kind and unshaking, even without beginning.[309]

and again

> It was not nor will it be all together but it is alone.[310]

and again

> For being abuts onto being.[311]

But if Parmenides says 'all together' and 'being abuts onto being' then he recognised that plurality exists among the things that are. But since their theory was not about physical things, but about intelligible things, for this reason they described them thus, because they were attending to the unification among intelligible things and to their immutability.[312]

<5.1.2 Textual analysis and exegesis, 186a22-b14>

> **186a22** And the same manner of arguments applies to Parmenides too, even if there are some other exclusive arguments.

That is, we shall refute Parmenides as well, when he says that being is one and motionless, by the same manner of arguments <sc. as those used against Melissus>.[313] For he <sc. Parmenides> also, in much the same way <sc. as Melissus>,[314] both adopts false premisses and argues fallaciously. Now if we are to use the same manner of arguments against Parmenides as well, it is clear that we shall resolve his arguments by a twofold refutation: (a) that they are false and (b) that they are fallacious. He will therefore be refuted by the same manner of arguments as Melissus, and by other exclusive ones. But people say that Aristotle wrote a book of his own exclusively against Parmenides' opinion, to which he is alluding here when he says 'even if there are some other exclusive arguments'.[315]

<5.1.2.1 Two problems with Parmenides' argument>

> **186a23** And the solution is (a) that it is false and (b) that it does not follow, [false in that it takes 'being' to be said simply, when it is said in many ways, but invalid …].

We shall resolve Parmenides' argument in two ways, Aristotle is saying: firstly on the grounds that he has adopted false premisses; and secondly on the grounds that, even so, this conclusion that he favours does not follow from those premisses. For indeed, Aristotle said from the beginning 66,1 that both Melissus and Parmenides 'adopt falsities and reason fallaciously'.[316] The falsities he <sc. Parmenides> adopts are the premisses, in that he takes the word 'being' as having a single meaning in the minor premiss, when in fact it has multiple meanings. For 'simply' here is equivalent to 'in one way', and that is clear from the fact that it is contrasted with 'in several ways': for he says 'in that it takes "being" to be said *simply*' (equivalent to 'in one way') 'when it is said in many ways'. But our view has been that the major premiss is also false.[317]

Such is the falsity in the premisses. But Aristotle does not seem to take the invalidity to be due to the conversion.[318] The text goes as follows:

> **186a25** But invalid in that if white things were taken to be unique on the grounds that 'white' has a single meaning, none the less white things would be a plurality and not one.

That is, even if we allow him to take the minor premiss as true and <grant> that 'being' signifies some one thing, e.g. white, it will be true to say that what is apart from being is not-being (i.e. what is apart from white is not-white) but the major premiss, which says that not-being is nothing (i.e. what is not-white is nothing), will no longer fit in. For what is not-white is not being, because 'being' signifies just one thing, namely white; however it will not at the same time also be *nothing*, since we too, when we say that matter is something else, besides all the things that have being – all the formed things, that is – because it underlies all of them, nevertheless do not say that it is *nothing*. So with the major premiss eliminated nothing will be inferred. For nothing follows from a single premiss.

Furthermore, even if we grant that being is one, and that it is white, none the less even so the things that have being will be a plurality. For the white things are several. The white can only be one either in genus or in species, but if the white is one in that way, clearly white things are several and not one; for the genus is a collective of more than one thing, and the species <likewise>. We shall certainly not take 'the white' as one individual; for that is impossible, Aristotle says. For if it were one individual it would have to be either one as what is continuous or as the same in specification, but both are impossible: the white is not one continuum (for they <sc. white things> are plainly distinct from each other); nor are they the same in specification. For the specification of white is different from the specification of the underlying recipient of white.

67,1 But let us see how Aristotle puts it in the text itself.

186a28-9 For being white will be a different thing from being the recipient.

Having said that even if 'white' has a single meaning 'none the less the white things are several',[319] because they will be one either in genus or in species, Aristotle adds 'For the white will not be one in virtue of continuity, nor in specification'.[320] That is equivalent to saying 'for the white will not be one as an individual', but the individual was said to be one either as continuous or as indivisible or as the same in specification.[321] Given, therefore, that nothing in the way of indivisibility extends to it, he passes over that at the moment. So he divides the <notion of> individual into two, and shows that in neither case is it possible to say that being (that is the white – 'white' is now serving to signify 'being') is one. Firstly he says that it cannot be one as being the same in specification:

> **186a28-9** For being white will be a different thing from being the recipient

– that is, it is impossible to give a single definition of what is white; for what is white is a body coloured in such and such a way, and one definition is given of the underlying substrate, another of the whiteness itself. White is not one of the things that subsists in its own right, but rather one that has its being in a body. So if it is impossible to give a single definition of the substrate and the whiteness, then what is white is not one by being the same in specification.

> **186a29** And there is nothing separate in addition to the white; for it is not in virtue of being separate, but in its being, that the white is different from that to which it belongs.

Given that Aristotle has said that the white is different from the substrate, because of this, lest one should think that there is a separate form of the white that is not in a substrate and for this reason the concept of the white is different from the substrate, he says that there is no other separate form (as Plato perhaps suggested) in addition to the white that has its being in the underlying body. For it is different from the substrate not because there is some separate white thing; rather, because it has a different essence (*ousia*), even if it failed ten thousand times to be separated, for this reason it takes a different definitional specification. Hence if what is white cannot be one in specification, and the same applies to speaking of being, then 68,1 neither is being one in that way.

But even if one were to suggest substance (*ousia*),[322] it would still be impossible for it to be one in specification. For the substance is receptive of the attributes, and they again take a different specification of their being. Hence being cannot in any way be one as being the same in specification.

> **186a31** But this is what Parmenides never saw,

namely that even if different things that take different specifications of their being are in the same thing, they are not one – due to the substrate – rather than several – due to their definitional specifications. It was seeing all the categories existing together that was the cause of Parmenides' going astray, for he thought that as a result being is also one.

> **186a32** So it is necessary to take 'being' not to signify just one thing, of which it is asserted, but also being-as-such (*hoper on*) and one as such (*hoper hen*).

Having shown that being is not one in virtue of being the same in specification, Aristotle now wants to show that it is not one in virtue of being continuous either. It has been said that if it is one as continuous, it is either substance alone, or attribute alone or a combination of the two.[323] At the moment Aristotle is showing that it cannot be just substance. 'It is necessary' therefore, he says, when listening sympathetically to their words, to take the term 'being' that they use 'not to signify just one thing', as they say, 'but also being-as-such and one as such'. 'Being-as-such'[324] – i.e. what strictly is, and that is substance – and again 'one as such' – i.e. what is strictly one, which is what is numerically one. For one would not call the attribute 'being', Aristotle says. That is what he meant when he said 'for the attribute is said of a substrate'.[325] That is, it has its being in something else. Hence in stricter usage one would call the substance being.

Next, having said this, Aristotle turns to demonstrating that they cannot be suggesting the attribute. For if that were what being is, he says, 'the thing to which being is attributed will not exist';[326] but it is attributed to substance; so that if the attribute alone exists, substance will not exist, for it will be other than being; but if substance is non-existent, then some entity will be a non-entity, for substance underlies the attribute, and hence substance is something. But <*ex*
69,1 *hypothesi*> the attribute alone exists, and hence substance does not exist. Thus the same thing both is and is not, which is impossible.

> **186b1** Being-as-such will not be something that belongs to another thing, then.

If it is impossible for the same thing to be both existent and non-existent, Aristotle is saying, but that occurs as a result of the fact that we assumed that being was attribute, then it is impossible for being to be an attribute of something else. For the attribute cannot be being, he says, 'unless "being" signifies a plurality of things',[327] so that both substance is a being and attribute too. For if 'being' signifies a single thing, the aforementioned absurdities will result. But the suggestion was that 'being' does signify one thing, so it is impossible that being is an attribute.

<5.1.2.2 Being as substance>

> **186b4** If, therefore, being-as-such is an attribute of nothing but of that something/why (*ti*) rather does being-as-such signify what-is than what-is-not?[328]

Having shown that they could not be suggesting that being is attribute, Aristotle goes back to the original claim that it could not be one substance either. There are two ways of punctuating this: either thus,

> 'If, therefore, being-as-such is an attribute of nothing, but rather something (*ti*) else is an attribute of it',[329]

and then having punctuated it so, to add,

> 'does being-as-such signify what-is or what-is-not?'[330]

Or instead one punctuates after 'but of that', and then continues

> 'why does being-as-such rather …'

and so on.[331] But the sense is this: if being-as-such cannot be an attribute of something, but something else is an attribute of that (i.e. of being-as-such) – that is, attribute <is an attribute> of substance – why does being-as-such (i.e. substance) rather signify what-is than what-is-not? In other words, substance will not be being any more than not-being. How this happens he then adds:

> **186b6** For if being-as-such and white are the same thing, but being for white is not being-as-such – for being cannot be an attribute belonging to it, for there is no being that is not being-as-such – then the white is not-being.

What Aristotle means is this: since being-as-such – i.e. substance – is also white (for we say that Socrates is white), but the white is not being-as-such (for only substance is being-as-such) then being-as-such is not being-as-such. So the same thing both is and is not, which is impossible. For if it is true to say that the substance is white, but the white is not being, then substance is not being. But then we were assuming that it was being, so the same thing is both being and not-being.

> **186b7** For being cannot be an attribute belonging to it; for there is no being that is not being-as-such. So the white is not being.

The white, Aristotle is saying, is not being. For being cannot belong to it; for all being is also directly being-as-such, but being-as-such is substance alone, and hence only substance is being. And if this is so, 'then the white is not being. Not in the sense that it is not-some-being-as-such, but not-being altogether'.[332] Since he has shown that the white is not being, because only substance is being, but what is not being includes both non-being *tout court*, and not being something (as Socrates is not something – for he is not Plato or Alcibiades), for this reason he says the white is not being, not in the sense of being not something, but 'not-being altogether', because that which alone is being, namely substance, also is not *a* being but being *tout court*. So that if substance is being *tout court* and the sole being, then the attribute is not-being *tout court*.

186b10 Hence being-as-such is not being.

This is the conclusion of the whole discussion. For if being-as-such is substance, and substance is white, but the white is not being, then being-as-such is not being. But Aristotle takes this up again and confirms it in what follows:

186b10 For it is true to say that it is white.

It is true to say that being-as-such is white, Aristotle is saying. For if being-as-such is substance, and it is true to say that the substance is white (for we say that the body is white), then it is true to say that being-as-such is white. But 'white' meant not-being (for only the substance is being). So that even being-as-such will be not-being.

186b11 Hence if 'white' too means 'being-as-such', then being means more than one thing.[333]

If it is absurd for being-as-such itself to be non-being as well, (because being-as-such is white, but white is not being), then it must be false that the white is not being, Aristotle is saying. So the white is being. But every being is immediately being-as-such: so that white also is being-as-such. But if this is the case, and substance is also being-as-such, then 'being' means not one thing but more than one.

71,1 **186b12** However being will not have any magnitude either, if being is being-as-such. For being is different for each of the parts.

Aristotle is taking the discussion onto a different topic. If substance alone is being-as-such, he is saying, then of necessity being-as-such cannot have magnitude either (for magnitude is a quantity, but quantity is an attribute; but attribute is not being), but if being-as-such is not going to have magnitude, it will not be finite or infinite either. Hence they <sc. the Eleatics> will undermine their own assumptions, by assuming that being means one thing, namely substance.

'For being is different for each of the parts' in the case of the substance that has size; Aristotle is referring to the substance itself and its quantity as if they were parts of the body. So, he says, even if these are one and the same thing as the substrate, yet being <that thing> is different for each of them. For the concept (*logos*) of substance is one thing and the concept of magnitude another. Hence if being-as-such is substance alone, then magnitude is not being; so substance will not have magnitude, lest the same thing should be both being and not-being. But if it does not have magnitude, it will be neither finite nor infinite; for these belong to quantity.

<5.2.1 Exposition and discussion: the structure of Aristotle's argument, 186b14-35>

186b14 However that being-as-such is divided into another being-as-such even in the specification (*logos*) is plain.[334]

Aristotle divided what is strictly one – I mean one numerically – into what is continuous and what is the same in specification (*logos*), and he said that if being were one in virtue of continuity, either it would be substance or attribute (for if it is a combination of both, evidently things are several). Then he showed that not even if it were substance would it be alone, since substance is receptive of attributes and it is true to say that the substance is white, but with the white not-being, substance too is not-being – yet it was supposed to be being, so the same thing will be both being and not-being. Given that the absurdity was drawn from the suggestion that substance was receptive of attributes, someone might say that perhaps there is no need at all to suggest that the attributes exist, but only substance exists. For this reason Aristotle now attempts to show that, even if we assume that being is substance, and that substance is numerically one and that the attributes do not belong to it at all, even so none the less things will be shown to be several and not one. In effect he says this: Are you teaching us about this substance or not? If you are not teaching us anything, but you expect us to go along with your propositions, we shall not be persuaded by you. But if you are teaching, either the teaching is entirely linguistic (*di' onomatos*), or it is conceptual (*dia logou*). If the teaching is linguistic, the language you are using is not conveying the nature of the items to us; for anyone who wants to know about the nature of a human being will not know anything further from the phoneme 'human'. And besides, if the words for things were sufficient to convey their reality (*ousia*), all of us humans would be equally knowledgeable, and we should all alike know what was signified by each of the words. But if you are teaching us conceptually, it is either entirely by means of a definition or by means of a description. And in general all education and all learning takes place by three methods: either (a) by negation (Plato said that lessons about god and about prime matter are learnt this way: that god is neither this nor this, but is 'above these',[335] and similarly matter is inferior to these),[336] or (b) by analogy (a method that Plato uses for god and matter again, saying that the relationship that the sun has to all the things in the world is the relationship that god has absolutely to all reality,[337] and the relationship that the bronze has to the bronze products is the relationship that matter has to all material things),[338] or (c) by definitions. But every definition and description is constructed out of a number of words, namely nouns and verbs.[339] These several words, therefore, either refer to one and the same thing or several various things. But if they refer to one and the same thing, then the teaching is linguistic again (for what difference does it make whether I indicate the object by the word 'sword' or 'knife' or 'blade'[340] for we refer to the same object by each of them), but if the expressions incorporated into the definition refer to several different things, they refer either (a) to several substances or (b) to several attributes; so if they refer to several substances, *ipso facto* the things that have being are plural (for substances are plural, and substances are things that have being); but if they refer to attributes, either they are attributes of being itself, or of something else, and they are either separable or inseparable.[341]

Aristotle therefore demonstrates that none of the hypotheses can be correct, but for the purpose of this demonstration he adopts four axioms:

73,1 First, that a separable attribute can also be detached from the subject;[342]

Second, that, given that some inseparable attributes belong to a certain definite kind of thing alone (as odd and even belong to number, straight and curved to line), while others belong to more than one (for black is an inseparable attribute of the Ethiopian, but also belongs to the raven and ebony and many other things), the subject is incorporated into the definition of an inseparable attribute belonging to one definite kind of thing,[343] – which is also one of the things that are signified *per se* (as nose <is incorporated> into the definition of snubness; for we say that snubness is concavity in a nose, and even is a number that divides in half);[344]

Third, that the parts are incorporated into the definition of the whole, but not *vice versa*. For example in defining human as 'rational mortal animal' we incorporate such parts as there are of a human, but in defining animal or rational we no longer incorporate human in the definition of any of these.

The fourth <axiom>, which Aristotle does not set out here – but he is going to mention it subsequently[345] – that the whole must belong to the same things to which the parts belong, and the reverse also, that the parts must belong to the same things as the whole. These are the axioms.

So Aristotle shows that the items incorporated in the definition cannot be attributes of the entity[346] in such a way as to be capable of being separated. As an illustration, let the entity[347] be one human being. The definition of this is rational two-footed animal. If rational and two-footed and animal are separable attributes of the human being, since a separable attribute is one that can be separated from the subject and not always be in it, it is clear that the human being will at some time be neither an animal, nor two-footed, nor rational. Hence these would not be definitive of the human being; for the things that are definitive of it are characteristic of it and always belong to it.

Hence it is impossible that the things incorporated into its definition should be attributes separable from the entity itself;[348] and the earlier premiss,[349] saying that the separable attributes do not always apply to the subject, has assisted us in this respect.

But if animal and two-footed are to be attributes inseparable from human being, either they are attributes of human being alone (like even <is an attribute> of number <alone>) or of other things as well, like black belongs not only to the Ethiopian but to ebony as well. If the items incorporated into the definition of the entity are attributes of it alone, then the subject of those attributes, that is human being, ought to be incorporated into the definition of each and every one of them. For the human being is the subject of them, on the hypothesis that says that these attributes are inseparable from it. But it is impossible for human being to be incorporated into the definition of animal or two-footed, for the whole is never incorporated into the definition of the part, and animal and two-footed are each a part, and the human being is a whole.[350] This is clear both from what is obvious and also from the following consideration. For just as the definition identifies the whole thing, so, plainly, the parts of the definition identify the parts of the thing; but if the subject cannot be incorporated into the definition of those parts, then neither can the items incorporated into the definition of the entity be attributes of that kind.[351] For since the things incorporated into the definition are parts of it, if the subject were incorporated into the definition of each one, then the whole would be part of the part, if it is incorporated into the definition of the part.

Hence the items incorporated into the definition of the entity cannot be inseparable attributes of it, as things that belong to it alone. This is demonstrated by means of the second and third premisses,[352] the one saying that the subject is incorporated into the definition of *per se* attributes; and the one saying that the whole is not incorporated into the definition of the part. What is *per se* is twofold:[353] (a) what incorporates the subject in its definition and (b) what is incorporated into the definition of the subject (rational, two-footed, and the parts of substances in general are of this sort); but Aristotle not unreasonably mentioned only the first kind. For this is the only one that applies to attributes, and the rest is said of substances (for it <sc. the *per se* attribute> is a part of the substance);[354] and should anyone suggest this is what we require, he would be granting that substance is divisible into substances. Thus it is clear from what has been said that the items incorporated into the definition of the entity cannot be attributes of it that are either separable or inseparable.

Perhaps, though, they might be attributes of something else. But if so, the whole would also be an attribute of the same thing; for what the parts belong to, the whole belongs to as well, and *vice versa*. But it is impossible for the entity to be an attribute of something else; so neither can it be that the items incorporated into its definition are attributes of another thing. This is demonstrated by means of the fourth premiss,[355] which says that the whole belongs to that to which the parts belong.

But suppose someone were to say that providing all the parts were attributes of one thing then quite rightly the whole would be an attribute of that same thing too, but if one part belonged to one and another to another (e.g. the 'animal' to one thing and the 'two-footed' to another; for what is two-footed is not *ipso facto* human) then no longer <is it so>, my reply is that in so far as I am taking the individual human being, clearly I shall also be taking its own 'animal' and 'two-footed'. But if this is the case, the whole human being will be split into parts, in such a way that whatever the 'two-footed'
75,1 belongs to, to that will also belong the part of the whole, and whatever all the parts belong to, to that will also belong the whole. Hence the things incorporated into the definition cannot be attributes of something else, either.[356] Thus they cannot be attributes at all. But if they are substances, then the entities are several and not one; for the entity is split up into entities that differ in form.

<5.2.2 Textual analysis and exegesis, 186b14-35>

186b14 However that being-as-such is divided into another being-as-such even in the specification (*logos*) is plain.

Aristotle has just said that 'being will not have any magnitude either'.[357] For if being is one, and this <one thing> is substance, evidently it will not have any magnitude, for that is <a type> of quantity. For if it does have magnitude, since the concept (*logos*) of quantity is different from that of substance, things will be more than one; for if someone suggested that quantity belonged to substance, but was not being, it would follow that the same thing was both being and not-being. It must be, therefore, that if substance is being, it has neither magnitude nor any other of the attributes. And on the one hand, it follows from this reasoning that they contradict their own hypotheses (for in this way it will neither be infinite nor finite), except that he now shows that even if we suggest that this is what they are saying, that substance is by itself alone without attributes, even so he shows that things will be several none the less. For this substance is divided into other substances in virtue of the definitional specification, even if these substances are the same as regards their underlying substrate. For human, despite being one in this very respect of being human, is divided into the animal and the rational and the mortal, all of which are different from each other in accordance with the definitional specification. One should therefore either punctuate after 'even in the specification (*logos*)' so that it should be 'that substance is divided into other substances that differ from each other in the definitional specification, even if not in the underlying substrate', or one should punctuate after 'into another being-as-such' and then take the next bit 'is clear even in the specification' to mean 'can be shown in the specification': i.e. 'that substance is divided into other substances, given that this is evident from what is obvious but also shown by specification (*logos*)'.

186b17 For if it is not a being-as-such, it will be attributes.

If the things into which human being is divided by the definitional specification are not substances, they must be attributes. But if so, they must be attributes belonging either (a) to it itself or (b) to another thing, and be either (a) separable or (b) inseparable, but both are impossible.

186b18 And this is what is said to be an attribute – either what admits of belonging and not belonging … 76,1

This is the first axiom, that a separable attribute does not always <belong>, but there are also times when it does not belong.

186b19 … or that in whose specification the subject of which it is attribute belongs.

The second axiom, to the effect that the subject is incorporated into the definition of the inseparable attribute – not of every inseparable attribute, though, but of the one that belongs *per se*. For snubness belongs *per se* to the nose.

186b21 … e.g. sitting as a separable <attribute>, …

An example of the first sort; for sitting belongs to a human being not always, but sometimes and sometimes not.

186b22 … but the specification of nose belongs in the snub.

An example of the second sort.

186b23 Furthermore the specification of the whole does not belong in the specifications of the things that are contained in its definitional specification or of which it is composed.

This is the third axiom, saying that the part is incorporated into the definition of the whole, but not the whole into the definition of the part. Aristotle says 'the things that are contained in the definitional specification', and then adds 'or of which it is composed'; for as he said in the *Posterior Analytics*, it is not possible to give a definition of everything – not of the most generic classes, nor of individuals; for if the definition is composed of genera and component differentiae, but it is not possible to take a class of the most generic classes, nor of the individuals (for genera are classes of species, not of individuals), clearly one could not give a definition of these things. For there would not be any common genus or species of Socrates, Plato, and Alcibiades, if each has its own characteristic collection of attributes, and it is not possible to predicate something synonymously of these collections. For whenever I say that Socrates is an animal or a human being, I do not predicate animal of the collection of attributes that characterises Socrates, except in a manner contrary to nature (a predication is said to be contrary to nature when we predicate the substance of the attribute, as when I say that this white <thing> is a man; for I am predicating in accordance with nature when I predicate the attribute of the substance)[358] – rather
77,1 I say that animal belongs to the particular human being, and not just any animal but the particular one belonging to that very thing. For Socrates is not animal in its generic sense in which it is predicated of all in common. In the same way we also say that 'human being' is predicated of him – again the <'human being'> that belongs to him, that is the one composed of this 'animal' and this 'rational'.

Given, as I said,[359] that not everything is indicated by a definition but there are some things that are indicated by a descriptive account as well, it is for this reason that Aristotle says 'or of which it is composed'. For neither does each of the things incorporated into the definition incorporate the whole into its own individual definition (e.g. 'animal' incorporating 'human' or 'horse' – into the definition of which it is incorporated – into its definition), nor does each of the four elements, out of which the composite body is made up, incorporate the composite body into its own individual definition,[360] nor indeed <is> the bed <incorporated> into the definition of the foot of the bed, nor Socrates into the definition of the hand.[361] But notice that he does not introduce the fourth axiom at this point,[362] but he will add it in due course.

186b26 If therefore these things work in this way and two-footed is an attribute of human being, it must necessarily be separable, in such a way that not being two-footed would be compatible with human being, or else the specification (*logos*) of human will be in the specification (*logos*) of two-footed.

If the axioms we have adopted are as stated, Aristotle is saying, and the parts of the definition of human being are attributes of human being, either they are separable attributes of it and it is possible for a human being sometimes to be not two-footed or animal or rational, or they are inseparable and we shall incorporate human being into the definition of two-footed or animal or rational. But both are impossible; for human being is always two-footed and rational and the wholes are not incorporated into the definition of the parts, but rather the reverse, the parts <are incorporated> into the definition of the wholes.

The next bit:

186b28 it must necessarily be separable ... or else the specification of human will be in the specification of two-footed.

i.e. it must be either a separable attribute of human, or inseparable.

185b31 But if two-footed and animal are attributes of something else, and each is not a being-as-such, then human being would also be one of the things that is an attribute of something else.

Having shown that it cannot be the case that the parts of the definition are attributes belonging to the thing itself whose definition it is, Aristotle wants to show that they cannot be attributes of 78,1 something else either. For if the parts of the definition are not substances, he says, (that is what 'and each is not a being-as-such' means), but if they are attributes of some other thing, since the whole also belongs to that to which the parts belong, the whole would also be an attribute of something else. Hence human being, and substance in general, will be one of the things that is an attribute of something, which is absurd and impossible, that substance should be attribute.

186b33 But let being-as-such be what is not an attribute of anything.

Since he has said that 'human being will also be one of the things that is an attribute of something else' and he has added to this that that account is absurd, for this reason Aristotle says that this is to be laid down as an agreement on our part – since it is also testified by what is self-evident – that substance is not one of the things that is an attribute of another thing, but is one of the things that subsists *per se*.

> **186b34** And that of which both [and each] is said, let the composite of them be also said of it.[363]

Since Aristotle adopted the view that the whole is an attribute of that of which the parts are attributes, without having posited it in advance,[364] for that reason he now sets out that thesis and says 'let this serve as an axiom for us, agreed on the basis of self-evidence, that that of which all the parts are predicated is also what the composite of them all is predicated of'. For if the whole is not a different thing from the parts, necessarily the whole must also belong to that to which all the parts belong. But if the text has the words 'and each'[365] it is to be interpreted thus: the whole composed of both parts will also be predicated of that of which both parts jointly and each part severally are predicated. But if the text is 'and in general'[366] it is to be understood thus: and by a general account both are to be predicated of the same thing, the whole also predicated of that of which the parts are predicated and *vice versa*.

79,1 **186b35** So everything is made up of indivisibles.[367]

This is to be interpreted in two ways. Either 'of indivisibles' means of points – since what is properly one – I mean what is numerically one – was said to be threefold, either *qua* continuous or *qua* indivisible or *qua* the same in specification.[368] Aristotle has been engaged in a refutation on the grounds that it (a) cannot be <one> *qua* the same in specification, nor (b) can it be one as what is continuous in any way, neither (i) taking the continuous as being substance, nor (ii) taking it as attribute, nor (iii) taking it as the compound of the two, and in the case of substance neither (iv) substance receptive of attributes, nor (v) non-receptive. Now, not surprisingly, he introduces the last remaining way. For if what is properly one was not one *qua* continuous nor *qua* the same in specification, what remains is that it is a point and made of points. It is a point in virtue of being neither continuous nor the same in specification; it is made of points in virtue of its definitional specification necessarily admitting several terms that signify things that turn out to be parts of it.

This is one way of taking 'of indivisibles'. Alternatively 'of indivisibles' means 'of substances'. For since Aristotle has shown that it is impossible for the specification of human being, or of being, to be divided into attributes, it follows that the division of it is into substances, given that the definition necessarily signifies a plurality of things. Hence the entity[369] would be composed solely of beings (*onta*), that is, solely of substances (*ousiai*). And in this way the entities are several, even if everything is not actually divided.

But if it is made solely of substances, it will never partake of quantity, but if that is so, it will not be either finite or infinite. And thus he was right to say 'everything' (*pan*) and not 'being' (*to on*), making 'everything' more striking since it is composed of divisible quantities. 'Everything', he says, 'which you said was one, will even so be composed of a number of substances, and would no longer be one, but several'.

<5.3.1 Exposition and discussion: on Parmenides and Zeno, 187a1-10>

187a1 But some have capitulated to both arguments.

Having tested the Parmenidean theory, Aristotle now says that we are not the only ones to dispute with Parmenides, but others before us did so too, even if they did not make their reply as they ought. Then wishing to test their inadequate solutions to Parmenides' argument, he first sets out Parmenides' argument itself – not the one he set out previously but the second one.[370] Parmenides' second argument is as follows: 80,1

> If being signifies one thing and contradictions are not simultaneously true, then being is one.

The two hypotheses in the argument included in place of one in the argument of the antecedent are 'if being signifies one thing' and 'contradictions are not simultaneously true', and the consequent is 'then being is one'. And this is no wonder, for it is frequently the case that the consequent follows not from just one hypothesis, but two: e.g. if I say 'if the soul is immortal, the subterranean regions are reform prisons' the consequent does not follow necessarily from the antecedent; for what if the soul were immortal but there were no such thing as providence? But if I combine the two hypotheses 'if the soul is immortal[371] and if there is such a thing as providence', then it will follow necessarily that the subterranean regions are reform prisons. But it would not follow from the second hypothesis alone either, for even if there were providence, but the soul was not immortal, it is not the case that the subterranean regions are reform prisons, since the thing that is judged does not exist. So just as here the consequent follows from the two hypotheses, but no longer follows from just one of the two, so it is with Parmenides' argument. For even if 'being' does signify one thing, but contradictions are simultaneously true, it no longer follows that being is one (for if contradictions are simultaneously true, plainly not-being is there as well; but if so, being is no longer one). If, on the other hand again, contradictions are not simultaneously true, but 'being' were to signify not one thing but more than one, it is clear that it will not follow that being is one. But if we combine both together, the one saying that being signifies one thing and the one saying that contradictions are not simultaneously true, it will follow validly, from both, that being is one. Such is Parmenides' argument affirming that being is one.

Zeno, the pupil of Parmenides, arguing in support of his teacher, affirmed that being is both one and necessarily motionless, and he affirmed these conclusions on the basis of the infinite division of continua. For if being were not one and indivisible, but were divided into more than one thing, nothing would be properly one (for if the continuum were divided, it would be divisible to infinity), but if nothing is properly one, neither are they several if the several are composed of several ones. So it is impossible for being to be divided into several; so it is uniquely one.

Or else as follows: if being were not one and indivisible, Zeno says, neither would it be several. For the several are composed from several ones. Each one is therefore either (a) one and indivisible, or else (b) is itself divided into several. If, therefore, (a) each monad is one and indivisible, everything will be composed of atomic magnitudes; but if (b) these monads too are divided, we shall ask the same things again concerning each of the divided monads; and this goes on *ad infinitum*. Hence everything will be infinity times infinite, if things are several. But if that is absurd, then being is uniquely one, and it is not possible for things to be several. For it is necessary to cut each of the monads an infinite number of times, which is absurd.

To show that this one is also motionless Zeno used the following sort of argument: if something traverses this finite line, it is entirely necessary that before it traverses the whole, it must traverse half, and before it traverses half of the whole line, it must first traverse a quarter, and before the quarter an eighth, and so on *ad infinitum*; for a continuum is infinitely divisible. So it must be the case that if something traverses a finite line, some infinitely many magnitudes are traversed first. But if so, and all motion occurs in a finite time (for nothing is moved in infinite time) then the infinitely many magnitudes will be traversed in a finite time, which is impossible. But what is infinite is wholly untraversible.[372]

In response to these arguments – that of Parmenides and that of Zeno – our predecessors capitulated, Aristotle says. He says that they capitulated because they put up too soft a resistance against the arguments – too soft a resistance because they fought against the conclusion while conceding the premisses; they were not tough enough to undo the arguments, but they struggled against the conclusion. They granted that 'being' means one thing, and that contradictions are not simultaneously true, but they did not go on to accept the conclusion, that being is one. They had not unravelled the arguments, but had reconstructed them to the opposite effect in such a way as to lead the discussion to a new impasse.

Both Alexander and Themistius say that Aristotle directs these allusions at Plato.[373] For Plato, they say, in the *Sophist*,[374] having adopted the hypothesis that there exists *not-being in general*, which escapes the nature of being, refutes the claim that all things are one by saying this: 'If all being is one, not-being will not exist. But indeed there is not-being. Therefore not all being is one'.[375] But how come Plato did not go on to accept that all things are also one, despite granting that being signifies one thing? Because he thought that by 'being' they <sc. the Eleatics> meant the really real – I mean the idea and what is intelligible – which he also claimed was the sole reality and besides it everything was non-being, and that it was one. But even if he also said that it was one and the only reality, he did not go on to say that everything besides it did not exist; for not-being also exists – that is the shapeless matter, which although quite different from being, nevertheless also exists. So he said that everything is composed out of both *being*, which he also granted was one, and also out of *not-being*, which he took to be matter. Hence he agreed with the premiss that states that being signifies one thing, but he did not go on to agree that absolutely everything whatever that exists is one, which is what the conclusion of the argument aimed at.

Aristotle upbraids Plato, then, for being soft in his resistance to Parmenides' reasoning, and this is because, having agreed that being signifies one thing, he does not go on to agree that everything that exists is one, on account of introducing another nature, which although he said it was entirely different from being, nevertheless he said that it also exists. This is no different from falling into contradiction; for having agreed that being is one, he goes back and says that things are several on account of there being not-being as well, which is entirely different from being. But how does this differ from saying that being is one and not one? For he does not escape the contradiction by saying that the former properly is and the latter not properly, without in fact falling even further into contradiction: for by making differences among beings he is plainly granting them a number greater than one.

For this reason, then, Aristotle upbraids Plato and says that there was a possibility, even granting that the nature of being was one, of showing that things were plural in this way, not by positing not-being *tout court*, which is wholly other than being (for that, as I said, is nothing else but falling into contradiction), but by positing what-is-not-something – which is not other than being but has its being in being itself – I am referring to what is-not in respect of otherness, which Plato himself says is not inferior to being. For even if it is inferred from Parmenides' argument that being is one, it is not

necessary to accept the 'one in number', for this does not follow from the premisses. For the claim that being has a single meaning is not that it signifies *numerically* one thing – nor are we going to grant that – but <that it signifies> *generically* one thing; hence the conclusion will also infer what is *generically* one. But if being is generically one
83,1 – as if it were animal or substance perhaps[376] – clearly it includes within it not-being as well – not not-being *tout court* but not-being-something, which is also an entity. For the human being is being and not-being: for he is not a horse or ass. But if not-being also exists, then being is plural, and this is not to fall into a contradiction. For we are not introducing the not-being that is totally deprived of being, while saying that being is one and that there is not anything else besides that. For this is nothing else but saying that the same thing is and is not. Rather <we are introducing> the *not-being that is in respect of otherness*, which is also being. Hence the same thing will be being and not-being in different ways – and not one and the same thing both being and not being in the same respect: he is not-being not *qua* human, but *qua* not being a horse. But on the other hand to say that there is something that is entirely different from being, and then to count it in with the beings and to say that there is a plurality of 'beings', is blatantly to make one and the same thing both be and not-be in the same respect. Otherwise one ought immediately to have resisted the premiss that stated that being signified one thing, and to have distinguished how many senses 'one' has, i.e. either nominally one or really one;[377] and if really one, either generically or specifically or numerically; and if numerically, either as continuous or as indivisible or as the same in specification; and to have demonstrated that on none of these hypotheses is Parmenides' argument sound, just as Aristotle himself did.[378]

This in response to Parmenides' arguments. But in response to Zeno's argument[379] they say[380] that Xenocrates reacted by supposing that division of magnitudes does not go on to infinity; for the line, when cut, comes to a halt at uncuttable lines. Xenocrates, too, was unaware that he had fallen into contradiction as a result of thinking he was avoiding a contradiction, for it is not impossible for the same thing to be both one and several, and such a case is not a contradiction if the one is potential and the other actual; but to make the same thing both a line and indivisible is blatantly to make the line not a line and the magnitude not a magnitude, if magnitude is indeed divisible to infinity.[381]

<5.3.2 Textual analysis and exegesis, 187a1-10>

187a1 But some have capitulated to both arguments …

By 'both' he means those of Parmenides and those of Zeno his pupil. To the arguments of the teacher, Plato the teacher <of Aristotle

<capitulates>; to the arguments of the pupil Zeno, Xenocrates pupil of Plato <capitulates>.[382]

187a1 … (a) to the argument that all things are one, if being means one thing, by saying that not-being exists … 84,1

They capitulated to Parmenides' argument establishing that all things are one, by accepting the premiss that being signifies one thing, when they ought immediately to have resisted the premiss and enquired in what sense they <sc. the Eleatics> were taking it that being signifies one thing, and <when they ought> to have distinguished how many senses 'one' has, and to have shown that on no sense of 'one' does their argument proceed correctly: which is what we have done. But those who went along with this premiss, fought against the conclusion, which necessarily follows, by suggesting that not-being also exists. But we have reported Plato's notion, to the effect that since he understood them to refer to the intelligible by 'being', he said that while matter does exist, yet it is not being – or rather all perceptible things as well – for this reason he granted both that being is one, and that it is not the case that all things are one, due to there being not-being as well.[383]

187a2 … and (b) to the argument from dichotomy, by making indivisible magnitudes.

– in that they also wrongly capitulated to the Zenonian paradox whereby he established that being is one and motionless on the basis of the infinite division of magnitudes, by mistakenly supposing that magnitudes are not infinitely divisible. For they granted that if magnitudes were infinitely divisible, there would be no motion, nor anything that is properly one, and hence no plurality either, given that a number is composed of several monads. Hence Xenocrates abolished the infinite divisibility of magnitudes. Aristotle meanwhile concedes to Xenocrates in lots of ways that he is right, but nevertheless he resolves the paradox in just one way in the passage in which he also says this – that the division of magnitudes continues potentially *ad infinitum*, but not in actuality (for an actual infinity cannot occur), so that motion traverses infinite magnitudes in potentiality, but finite ones in actuality; for motion does not occur at a point, but over a finite magnitude.

Aristotle calls Zeno's enterprise 'dichotomy', and it is clear why – because this is how he constructed the proof, by always cutting the 85,1
remaining magnitude in two.

187a3 But it is also plain that this is not true, that if being signifies one thing and contradictions cannot be simultaneously

> true, there will be no such thing as not-being; for nothing prevents not-being – not from being *tout court*, but from not being a certain particular thing.

Plato, Aristotle says, was not correct in supposing that Parmenides' argument entailed eliminating not-being wholly and that it could not exist in any way. For the conclusion that follows upon the entire line of argument is that being is one, not that not-being in no way is. For given that Plato called matter 'not-being' and said that it was excluded from the nature of being, but followed Parmenides' theories in asserting that there is nothing that is not of the nature of being, hence, Aristotle says, Plato concluded that it follows from the Parmenidean argument that not-being in no way is. But this is not true, Aristotle says; for there is nothing to prevent it being the case both that this argument is true – the one that proves that being is one, I mean – and also that not-being exists. But when I say not-being in this case I do not mean the *tout court* not-being – which is entirely different from being – but not-being-something, which exists in reality, i.e. *the not-being that is in respect of otherness*.

Accordingly, the sentence 'this is not true, that if being signifies one thing and contradictions cannot be simultaneously true …' is an alternative way of saying 'it is not true that it follows from the entire argument that there is no not-being in any way'. Aristotle has put in the premisses in place of the whole argument. The next words 'for nothing prevents not-being' do not refer to not-being *tout court*, but to not-being-something.

> **187a6** But (*de*) it is absurd to say that besides being itself, supposing there is nothing else, all things will be one.

By these words Aristotle proves that it does not follow from Parmenides' argument that not-being does not exist. We are to understand the connective 'but' (*de*) as a substitute for 'for' (*gar*) – 'for to say that if there were, besides being, nothing else that is entirely other than the substance of being, then necessarily all things are one, – as Plato was forced to posit not-being *tout court*, on the basis that this did follow, in order to show that things were several that way – that is absurd', he says.

> **187a8-9** For who understands 'being itself' unless it is a being-as-such?

– Aristotle says. In other words, who, on hearing them say 'being' will not think that they mean being-as-such – i.e. what properly is, by which I mean substance? But if that is being, there is nothing to prevent it from being several, as we have said. For we shall understand being to be one generically and not numerically. But if so, clearly not-being-something is in this <i.e. in the class of being, understood generically>; for each of the species under the genus, and each of the individuals under the species, *is not* what the other is. So it is not correct to think that Parmenides' argument entails eliminating not-being.

On the other hand the following proposal of my own is also possible – that positing not-being was not essential either, to refute Parmenides. For if it is possible to understand 'one' generically and not numerically, it is superfluous to want to show that there is a plurality of beings by positing not-being, when one could have shown that there is a plurality of beings directly by means of classification (*diairesis*). For the one is generic, not numerical. On this interpretation we shall not substitute 'for' (*gar*) for 'but' (*de*), because it is an initial '*de*'.[384]

Notes

1. i.e. *Historia Animalium*. Philoponus' list of the zoological works does not include *De Generatione Animalium*.

2. i.e. the *Parva Naturalia*.

3. In the first part of this paragraph Philoponus appears to accept without question Aristotle's division of things into (a) eternal things and (b) things involved in coming to be, and reports without comment that the *De Caelo* is about heavenly bodies that are eternal. Philoponus' own developed position rejects the eternity of the heavenly bodies. However, the present passage may be simply designed to classify *Aristotle's* work, not to engage with the issues.

4. i.e. 'natural <philosophy>'.

5. i.e. the Presocratic philosophers.

6. Democritus of Abdera, Presocratic philosopher, floruit *c.* 460-57 BC, famous (with his contemporary Leucippus) for the invention of ancient atomism. See KRS, ch. 15.

7. Anaxagoras of Clazomenae, Presocratic philosopher, approx 500-428 BC. See KRS, ch.12.

8. Homoiomeries (meaning things with similar parts) is a technical term, probably introduced by Aristotle, for the basic stuffs out of which the world and things in it are made in Anaxagoras' system. These are stuffs like flesh, bone, and the like, which when divided have parts that are like the whole (unlike, for example, a face or hand). Anaxagoras held that everyday things contain portions of all the available homoiomeries. See below, 24,24-25,4.

9. *Phys.* 4.6-9, 213a12-217b28.

10. *Phys.* 3.4-8, 202b30-208a23.

11. cf. above, 2,20.

12. Philoponus writes *ginôskein* (know); MSS of Aristotle read *gnôrizein* (discover).

13. *Metaph.* 1.1, 980a21.

14. *An. Post.* 1.1, 71a1.

15. *EN* 1.1, 1094a1.

16. *sullogismos*. Philoponus is here attempting to reformulate Aristotle's reasoning using Aristotelian syllogistic logic (following Aristotle's logical doctrines in, e.g., the *Prior Analytics*).

17. Theophrastus (*c.* 370-288 BC), pupil of Aristotle. His *On Nature* is not preserved (but see FHSG text 144 and commentary, on this passage and the parallel in Simplicius *in Phys.* 9,5-10). See also de Haas, 'Philoponus on Theophrastus on Composition in Nature', and Laks, 'Le début d'une physique: ordre, extension et nature des fragments 142-144A/B de Théophraste'. De Haas argues that lines 15-34 – marked within parentheses in the Greek by Vitelli – are an expansion by Philoponus and not paraphrased from Theophrastus.

18. *pneuma*. The physiological basis of vision forms the underlying substrate in which sight (the faculty) becomes actual. The account is not strictly Aristotelian but appears to correspond to the Stoic account of vision (see for example Iamblichus' explanation of the Stoic theory of vision, ap. Stobaeum 1,368,12-20, in L&S 53K). But compare Philoponus' *De Anima* commentary where he similarly invokes the idea that the underlying bodily equipment for the faculties of soul is *pneuma* (See e.g. *in DA* 19,35 etc.) Although this paragraph purports to be explaining Theophrastus' reading of Aristotle's argument, the illustration in terms of vision may be Philoponus' own.

19. i.e. the Presocratic Philosophers.

20. The claim that definitions exist only in thought (i.e. not as Platonic Forms) is repeated from line 19 above. It is not clear whether it is part of the report of Theophrastus' reasoning or an assertion made by Philoponus in his own voice.

21. This passage, along with 39,3-7, are important for understanding Philoponus' views on the relation between body, matter, and extension. It has been suggested that *in Phys.*, being an early work, shows Philoponus not yet holding his developed view (found explicitly in *Aet.*) that three-dimensionality is the ultimate substrate, or 'first subject' of body. But his position here is subtle. Here he claims that body is properly defined as simple, mere three-dimensionality, and that it becomes complex, with the addition of prime matter as a condition of its existence (*huparxis*). This already suggests a departure from Aristotle, in that matter is not an essential part of the definition of what body is. Later he will repeat that suggestion (38,21) and ascribe to Aristotle (but not to himself) the thought that body is a complex of matter and extension (39,6-9). See R. Sorabji, 'John Philoponus', 18-23; Verrycken, 'Development'.

22. Philoponus resumes the summary of Theophrastus' exposition by repeating the claim sketched above at lines 12-15, for which the intervening material offered a supporting argument.

23. Plato does not provide a systematic treatment of causation of this kind in any of his extant works. This kind of classification, assembling evidence from hints in a variety of Plato's works (see further below), belongs rather to the Neoplatonic tradition, i.e. typically re-interpreting and systematising Plato on Aristotelian lines.

24. Plato *Timaeus* (49A and *passim*).

25. cf. *Timaeus* 46C7 ff.

26. Note that Aristotle's four elements (earth, air, fire, water) are composed of matter and form. Matter is a substrate below the level of elements.

27. At this point Philoponus embarks on some detailed comments on the terminology of the first lemma. There is no suggestion of a change of gear from introductory discussion to detailed textual exegesis. These comments on the terminology for 'causes' are clearly intended to be of direct relevance to the discussion of causes and principles just given. The detailed exegesis of the lemmata of this section commences below at 6,29.

28. *Phys.* 1.2, 184b15.

29. *Phys.* 1.5, 189a9-10.

30. I am reading *mia men oun arkhê autê, tên hulên legôn*, as suggested by Vitelli in his apparatus (ad loc.) from Aristotle.

31. *Phys.* 1.7, 191a12-13.

32. Literally 'a pathmaking attitude with reason'.

33. Above, 3,25-4,6.

34. *Phaedo* 75D.

35. See above, 7,28. The non-uniform parts are the limbs and organs of an organic body; the uniform parts (*homoiomerê*) are materials like flesh, bone, and blood. The technical terms are Aristotle's and are often used in his account of Anaxagoras' theory (on which see 2,24; 24,24).

36. Above, 7,26-32.

37. Above, 4,3-4.

38. *Phys.* 184a10-16.

39. *An. Post.* 1.1. Aristotle refers to a distinction between (a) demonstrative (*apodeiktikê*, 71b20, or *sullogismos*, 71a5-11) and (b) inductive (*epagôgê*, 71a6). Cf. also *Top.* 105a10-19.

40. Or perhaps 'using a second rate type of demonstration'.

41. *tekmêriôdês.* Aristotle uses this term in the *Rhetoric* 1403a11. It is a favourite technical term among the commentators for this kind of inverted reasoning from what is familiar to what is prior and explanatory in nature. Philoponus uses it regularly in his commentary on the *Posterior Analytics.* Cf. also his *De Anima* commentary 31,15 (where he uses the same example of smoke and fire). Simplicius also uses it in explaining the lemma under discussion here (Simplicius *in Phys.* 15,24).

42. *Cael.* 2.11, 291b18 ff.; cf. *An. Post.* 78b5 ff.

43. It has been suggested that this passage shows Philoponus accepting Aristotle's doctrine of the fifth element (Evrard, 'Les convictions religieuses de Jean Philopon et la date de son Commentaire aux Météorologiques', 324-5; Verrycken, 'Development', 235). In fact Philoponus is merely illustrating the method used by Aristotle whereby he *aims* to demonstrate the shape of the moon; Philoponus suggests a way in which Aristotle might have done so more strictly in accordance with rules of priority, on the basis of Aristotle's own assumptions. Philoponus does not imply that it would then be a well-founded proof. But for further evidence that Philoponus accepts the doctrine of the fifth element see below 15,30 and 16,2-8.

44. *Metaph.* 7.3, 1029a10-26.

45. Aristotle regularly draws a distinction between what is clear to us and what is clear and more knowable in nature; cf. *Metaph.*7.3, 1029b3-8 which seems to make a similar point to *Physics* 1.1. Things 'clear to us' seem to be entities we meet in everyday experience; things more knowable 'in nature' are not directly available to perception but have a more fundamental role in explanation of the phenomena. It remains puzzling why Aristotle seems to suggest here that the universal is more familiar than the individual.

46. *sunkekhumena, Phys.* 184a22. This word is sometimes translated 'confused' or 'compounded'.

47. Philoponus' interpretation of what Aristotle means by *sunkekhumena.* This interpretation is supported by Aristotle's subsequent comments, *Phys.* 184a24-6.

48. cf. *EN* 7.3, 1147a25-6; *An. Post.* 2.19, 100a15-b5; *Top.* 156a3-7.

49. *Int.* ch. 7 distinguishes the universal and the individual, but it does not explicitly make the distinction Philoponus is introducing here between individual and particular.

50. Philoponus is making the distinction between 'a man' (the particular) and 'Socrates' (the individual). 'A man' applies to any individual man you care to mention, but 'Socrates' applies to one only. Hence 'a man' is more general or universal.

51. cf. *SE* 179a33-b4.

52. *Phys.* 184b12-14. Aristotle speaks of calling all men (*andres*) 'fathers' and all women (*gunaikes*) 'mothers' whereas Philoponus speaks of calling all human beings (*anthrôpoi*) 'fathers'.

53. The reference seems to be to universals that are features inherent in natural things.

54. *pro*, 'before' or 'in front of'. The reference seems to be to transcendent universals that are the subject of metaphysics.

55. *epi* (with dative), 'at' or 'over'. The reference seems to be to the results of inductive reasoning leading to generalisation.

56. See above 10,28-31.

57. Vitelli suggests that this section in square brackets should be excised.

58. Above, 12,3-4.

59. Advocates of the view that Philoponus is currently criticising (that Aristotle held that we are to start from the universal, the genus in the many) would need to be committed to the view that this universal is posterior and less clear in nature, in order to be consistent with Aristotle's claim that we proceed from what is less clear in nature to what is more clear in nature. Philoponus here allows that there is a sense in which the universal is posterior in nature.

60. 12,7.

61. 10,28-11,23.

62. Philoponus has not said precisely this before (but cf. 12,19-20).

63. Some manuscripts have the 'but', while others omit it so that the sentence reads grammatically despite the lacuna. Vitelli suggests supplementing as follows: '... prior for us, but it is necessary with a view to knowledge of the principles to begin from things that are prior and more clear for us, it is necessary also in the discussion ...'.

64. Philoponus' use of the word 'particulars', here and at lines 16 and 29, seems to be misleading and imprecise. His point appears to be that our procedure in gaining knowledge is to start from the particular (general) and work towards precise knowledge of individuals, whereas nature always starts from individuals and subsequently constructs universals, but never makes particulars. In both these cases where he says that nature makes the particulars, he should strictly have said 'individuals'; otherwise his claim that our procedure is the reverse of that of nature fails.

65. Again 'particular' seems to be used interchangeably with 'individual'.

66. Again 'particular' seems to be inaccurate terminology in place of 'individual'.

67. See above, 14,4-20.

68. Here (and below, 16,1-10) Philoponus appears to accept Aristotle's theory that the heavenly bodies are of a different kind of matter from sublunary things, and that they are eternal whereas earthly things are subject to coming to be. See Verrycken, 'Development', 235. Since it is possible to hold that the stuff of which the heavenly bodies are made differs from the stuff of natural things on earth (as the stuff of plants differs from the stuff of animals) while maintaining that the heavens were created, and Philoponus may at some stage have held this, the only significant commitment to Aristotelianism is in referring to the heavens as eternal or not involved in generation.

69. This is almost a repeat of the sentence at 15,30, but here perhaps it is to be taken as meaning that there is a different *prime matter* in the case of heavenly things, and not merely a different kind of matter of the same order, a fifth element on a par with the other four.

70. Above, 14,4-20 and 15,21.

71. Although the demiurgic imagery seems somewhat out of keeping with Aristotle, Philoponus follows Aristotle in treating nature as species-specific. There is no 'Nature', over and above the natures of the various things in nature, to be responsible for creating the matter that is common to them all. But see below, 17,2-3 and note ad loc.

72. Philoponus' denial that matter is made by nature is compatible either with an Aristotelian view of the eternity of the world, or with Philoponus' developed view on the divine creation of matter. When he says that matter 'was provided from the first' it is unclear whether he means it is without origin, or was created at the beginning of time.

73. Here Philoponus does refer to 'Nature' as a general thing, not species-specific. He does not suggest that he himself subscribes to such a view: in order to hold that formless matter was better known, one would have to hold that it was better known to some such general nature.

74. cf. *An. Post.* 1.2, 72a1-5.

75. *Phys.* 184a16-18.

76. *Phys.* 184a 23-5.

77. *An. Post.* 1.2, 72a1 (freely cited).

78. See above, 10,22-14,20.

79. The reference is not to any specific section of the *Meteorologica* but to the general method of treating particular physical topics there. Thunder is one of the topics in Book 2 of the *Meteorologica.*

80. Alternatively 'he says there that the reason <for this> is plainly the nature of the universals', or 'he says that the reason is that the nature of universals plainly goes that way'. The reference to the *Posterior Analytics* here is to the passage mentioned above (17,12-18,4): here Philoponus concludes his explanation of the apparent contradiction between the *Physics* and the *Analytics.*

81. *An. Pr.* 1.1-7.

82. *diairesis*, division, a technical term for a logical taxonomy or catalogue dividing a topic systematically into subdivisions, and subdivisions of those subdivisions.

83. *antiphasis* (contradiction, or negation, or complement). Philoponus means that the divisions in a scientific analysis are based on pairs of predicates of which one is the negation or complement of the other (f or not-f) forming a pair of exhaustive alternatives: everything must fall into one or other of the two classes, so that the classification is always inclusive and complete.

84. That is, it is not in the form 'x is f or not-f', as it would be if it said, for example, 'the principles are either one or not-one'. Philoponus takes it that 'either one or more' is logically equivalent to 'either one or not-one'.

85. Parmenides of Elea, early fifth century BC. His poem (in three parts, Prologue, Towards Truth, Towards Opinion) is partially preserved in quotations in the commentaries of Simplicius on Aristotle's *Physics* and *De Caelo.* I translate *pros* as 'towards' to capture the metaphor of paths of travel in Parmenides' poem. See KRS, ch. 8, and Guthrie, vol. 2.

86. Xenophanes of Colophon, *c.* 570-470 BC. Extracts from a number of his poems on nature, theology, and theory of knowledge are preserved. See KRS, ch. 5, and Guthrie, vol. 1.

87. The allusion is actually to *GC*, 325a18.

88. *hoi peri Parmenidên*, literally 'those around Parmenides'.

89. cf. *GC* 318b6; 330b14.

90. Or 'the commentators on Aristotle'.

91. *ontôs onta.*

92. *alêthôs onta.*

93. Or 'thinkables', matters of opinion, *doxasta.* The word is etymologically related to the word for 'opinion', *doxa.* See below 55,32.

94. Plato *Timaeus* 28A1-4. Philoponus attributes to Timaeus the words spoken by the character Timaeus in Plato's text.

95. Philoponus here subscribes to the Neoplatonic interpretation of Parmenides, which assumes that the *Towards Truth* describes the intelligible world of Neoplatonic thought, and the *Towards Opinion* refers to the perceptible world. Since the Neoplatonic view is that the intelligible world comprises a plurality of intelligible entities, this interpretation has difficulty coping with the radical monism of Parmenides' *Towards Truth*. Hence Philoponus attempts to explain the reference to 'one' in Parmenides as a recognition of the fact that the intelligibles are closer in derivation to the 'One' that is the source of all being in Neoplatonism. For modern discussion of the vexed question of the relation between *Towards Truth* and *Towards Opinion* see KRS, pp. 254-6, and e.g. Long, 'The Principles of Parmenides' Cosmogony', and Mourelatos, *The Route of Parmenides*.

96. I am omitting the word *apeiron*, entered in square brackets by Vitelli. If *apeiron* were retained the sentence would read: 'but Melissus, who was also talking about the same things, as I said, held that it was one and motionless and unlimited'.

97. Hippasus of Metapontum, fifth century BC, member of the Pythagorean school. See Barnes, *Early Greek Philosophy*, 214-5.

98. Heraclitus of Ephesus, fl. *c.* 500 BC. See Osborne, 'Heraclitus'. The claim that Heraclitus made fire a material element, in the same way that other natural philosophers conceived of this, is controversial.

99. Thales of Miletus, early sixth century BC; according to tradition the first Greek Philosopher. See Barnes, *Early Greek Philosophy*, 61-70.

100. Hippon, probably of Samos, or of Rhegium, Croton, or Metapontum, fifth century BC. See Guthrie, vol. 2, 354-8.

101. Diogenes of Apollonia, fl. *c.* 440 BC, a younger contemporary of Anaxagoras. See Guthrie, vol. 2, ch. 7; KRS, ch. 16.

102. Anaximenes, the third of the Milesian philosophers, fl. *c.* 546 BC. See KRS, ch. 4.

103. Anaximander, the second of the Milesian philosophers, *c.* 610-540 BC. See KRS, ch. 3. He is noted for having proposed a material stuff called 'the indefinite' (*to apeiron*), which has the character of no one of the familiar stuffs in the world.

104. Philoponus' objection to Anaximander here is almost certainly unjust since there is no good reason to think that Anaximander did envisage the indefinite principle having a definite density as Philoponus makes out. The suggestion that the indefinite principle was an intermediate between air and water or fire and air appears to be supported in a number of passages of Aristotle: *GC* 2.5, 332a19-25; cf. *GC* 2.1, 328b35; *Phys.* 1.6, 189b1; 3.4, 203a16; *Cael.* 3.5, 303b10; *Metaph.* 1.7, 988a30; 989a14; there is a problem with reconciling these passages with another passage in Aristotle, *Phys.* 1.4, 187a12-23, which implies that Anaximander was not the author of the theory of an intermediate. Philoponus evidently ascribed to Anaximander a theory of an intermediate, but recent scholarship generally rejects the testimony of Aristotle and his commentators on this point. See KRS, pp. 111-13.

105. *apeiron*. This was the term used by Anaximander for his primary stuff. Aristotle seems to take it to mean spatially boundless or inexhaustible in quantity (i.e. infinite), but it also means 'indefinite' or 'indeterminate', and it is possible that those implications were more important for Anaximander than the notion of infinity that interests Aristotle in the *Physics*.

106. This point is made by Aristotle at *Physics* 3.4, 203b15-20, where he seems to infer that the early philosophers took this as the reason for proposing an infinite principle. Aristotle himself denies that it is a valid reason, *Physics* 3.8, 208a8-11, on the grounds that the passing away of one thing can be the coming to be of another. This is the 'theory of recompense and the alteration of things one into another' that Philoponus now denies that they held. It was because they did not hold that theory that they had to assume that the principle must be inexhaustible, he suggests.

107. The claim that Anaximander did not attend to a notion of exchange or 'recompense' is surprising. Compare Simplicius on Anaximander, *in Phys.* 24,17: 'And the source of coming-to-be for existing things is that into which destruction, too, happens, "according to necessity; for they pay penalty and retribution to each other for their injustice according to the assessment of time", as he describes it in these rather poetical terms' (= DK 12A9 and 12B1, trans. KRS).

108. The reference is to Timaeus in Plato's dialogue the *Timaeus*. See particularly 27D-31B (god and form); 47E-48E (matter). The word for 'form' here is *idea*.

109. Empedocles of Acragas, Presocratic philosopher, *c.* 495-35 BC. Empedocles' cosmological speculations focused round an oscillating sequence in which a world composed of four elements and compounds of them alternates with a unified world called the Sphere. The forces of love and strife figure as key factors in the alternation between one and many. See KRS, ch. 10, and Inwood, *The Poem of Empedocles*.

110. The interpretation of Empedocles which follows is heavily coloured by Neoplatonism, particularly in importing the notion of perceptible and intelligible worlds. On the Neoplatonic interpretations of Empedocles see O'Brien, *Pour interpreter Empédocle* and also Osborne, *Rethinking Early Greek Philosophy*.

111. *sphairos*. The term is masculine in the Greek, unlike the normal term for sphere which is *sphaira*, feminine. The masculine form seems to be almost unique to Empedocles.

112. Plato *Timaeus* 36C.

113. Plato *Timaeus* 36C.

114. Empedocles, DK 31B115, 13-14. The text of line 13 differs here from that standardly adopted by the editors of Empedocles, which reads 'of these I now am one, a fugitive from god and a wanderer', based on readings from Hippolytus *Ref.* 7.29.14 and Plutarch *De Exilio* 607C. Philoponus' reading is replicated in his other commentaries (*in GC* 266,4; *in DA* 73,32) and in Asclepius *in Metaph.* 197,20. See Wright, *Empedocles: The Extant Fragments*, 138-9, 270-2.

115. Epicurus, 341-271 BC. Founder of Epicureanism. Epicurus revived atomism as a sophisticated physical theory. See L&S §§ 4-15.

116. Leucippus of Miletus, Presocratic philosopher associated closely in ideas with Democritus, fl. *c.* 440-35 BC. Together Leucippus and Democritus are the originators of ancient atomism in the Presocratic period. See KRS, ch. 15. For Democritus, see above 2,23.

117. On Anaxagoras see above 2,24.

118. On this technical term see above, notes to 2,24 and 8,5-6. I have translated it 'uniform parts' at 7,28 and 8,6.

119. i.e. the part (*meros*) is similar (*homoion*).

120. This observation is relevant because it completes the division into the various possible views regarding the number and nature of the principles: see above 21,9-13.

121. This term, *genos*, can mean 'kind' or 'genus'. I have translated it 'kind' below, 25,22; elsewhere I have sometimes translated it 'genus' (see for example

12,5 where the subject is universals). It seems that Philoponus may read this lemma as making a contrast between *genus* and *species*, prompted by Aristotle's choice of the terms *genos* and *eidos* (the latter here translated 'form' but often used for 'species').

122. This claim is repeated twice below (129,17; 203,15) though without specifying the *Posterior Analytics* as the location. See e.g. *An. Post.* 75b1 and 76a12 for the use of *genos* with respect to the subject matter of a science.

123. Most modern editors of Aristotle take *skhema* and *eidos* to mean different things, adding an additional 'differing', so as to offer just two distinct alternatives: 'differing in shape or differing (or even opposed) in form (species)'. As Philoponus goes on to show, he favours the view that 'differing in shape or form' is one alternative (Democritus'), with 'shape' and 'form' being synonyms, and then there is a further distinct alternative, not the view of Democritus, namely the suggestion that they are opposite – a view he attributes to Anaxagoras.

124. This sentence is ambiguous and hard to construe. It is not entirely clear what Democritus said, or whether Philoponus intends to attribute some particular phrase or formula to him.

125. i.e. it neatly completes the schematic division into exhaustive alternatives, by offering two ways in which one may have an infinite number of principles, either numerous different sorts under a single genus (Democritus), or with opposition between quite contrasting kinds (Anaxagoras). Philoponus did not include this subdivision in the classification as he sketched it above, 20-1.

126. At 184b22-4 Aristotle appears to equate the investigation of 'how many things there are' with the investigation of 'how many principles there are', so as to justify including thinkers who simply list the ingredients of the world, not the origins. Philoponus reads Aristotle as making excuses for his systematic classification by division, and for listing the various options for how many principles there might be, on the grounds that such a classification is implicit in the enquiries of the natural philosophers whose investigations he is collating.

127. i.e. metaphysics; see Aristotle *Metaph.* 6.1, 1026a23-32. For the term *anupothetos* see Plato *Republic* 510B; 511B.

128. 'The common science of all' mentioned by Aristotle at *Phys.* 185a2-3.

129. *Topics* 1.1, 100a1-3. The text differs in minor details from Aristotle's text. *Endoxa*, here translated 'ordinary opinions', are the received views, whether held by philosophers or common sense, that are worthy of consideration, and form the starting point of enquiry for Aristotle.

130. The geometry example is Aristotle's (*Phys.* 185a1) but Philoponus has filled out the nature of the principles that are under attack. The other examples of the particular sciences, and the hierarchy in which they are here presented, are not explicit in Aristotle.

131. Literally 'first philosopher', here and elsewhere.

132. Neither 'music' nor 'grammar' carries precisely the same sense in the Greek as it does normally in English. 'Music' covers the arts of the muses in general, including lyrics, poetry, and literature. 'Grammar' covers reading and writing, the various skills and arts connected with the written word. We should perhaps understand 'grammar' here to cover the art of reading literature aloud with the correct vowel quantities (recognising long and short vowels, but without requiring the theory of what accounts for the vowel being short in this word, say) and music as covering the art of composition in metre, which requires knowing both the metrical rules (where long and short are needed and why) and how to obtain the effect (i.e. which sequences of letters or parts of speech will generate that quantity for the vowel, and why). Thus music includes a branch of linguistics, as well as metrical theory (and also harmonics, as explained in the next lines, to the extent of knowing what length of string generates this musical interval, but not why that ratio has that effect).

133. I have retained Vitelli's text, which is the reading of t. Other manuscripts have a variety of readings, some of which suggest that there was a mention of *astronomia* in connection with the mathematical sciences, but no coherent text to that effect seems to be recoverable.

134. Or 'the principles of all <things>'.

135. I am omitting *anankê pasa* excluded in square brackets by Vitelli on the grounds that it has crept in from the later occurrence in line 11.

136. *Phys.* 185a4-5.

137. Motion (*kinêsis*) is used throughout this passage to refer to the broader notion we usually call 'change'.

138. *genesis*, here translated 'birth', has a broad meaning covering all forms of coming into being, generation, or creation.

139. Philoponus appears to change gear at this point from the general discussion of the structure of Aristotle's thought to details of particular expressions. I have therefore inserted a heading and read the following sentence as a lemma introducing the first section of detailed textual exegesis.

140. i.e. Philoponus is saying that Aristotle constructs a syllogism in the second figure (that is, a syllogism that takes the general form P is M, S is M, therefore S is P). Philoponus' example is a hypothetical syllogism in propositional logic, of the following form: if p then $\neg$q, if r then q, therefore if p then $\neg$r.

141. The wording '*ei houtôs hen*' (if one like that) might suggest that Philoponus was looking at the second occurrence of *houtôs hen* (one like that) at *Phys.* 185a5. However that belongs to the next lemma, below. More probably he refers to the occurrence of *houtôs hen* at 185a4, which needs to be read with the 'if' (*ei*) from earlier in the sentence.

142. *Topics* 1.11, 104b19-20. Aristotle there gives as examples (a) Antisthenes' view that contradiction is impossible; (b) Heraclitus' view that everything changes; (c) Melissus' view that being is one. Philoponus offers a slightly different range of examples here.

143. Zeno of Elea, born *c.* 490 BC, disciple of Parmenides. See KRS, ch. 9.

144. The view attributed to Anaxagoras here is presumably meant to apply to his physical theory, but is not a familiar description of Anaxagorean doctrine and it is not obvious how Philoponus supposed it to apply. In fact Philoponus' expression is almost identical to that used by Aristotle (in the *Topics* passage cited, 104b21) to describe not a view of Anaxagoras, but rather *Heraclitus'* view that everything changes, and Heraclitus is indeed a more obvious candidate for this doctrine.

145. Or: 'that the whole of reality is one individual human being'. The example is Aristotle's (*Phys.* 185a7, paraphrased with the addition of 'all' (or 'whole') and 'individual' by Philoponus); it is not clear whether anyone is supposed to have put forward such a thesis. Philoponus is perhaps trying to equate it with solipsism.

146. Aristotle speaks of the premisses of an argument as its 'matter' (e.g. *Phys.* 2.3, 195a18-19). Philoponus means that a sophistic argument is one that has true premisses but the reasoning is invalid, whereas an eristic argument has false premisses and invalid reasoning.

147. *hiptatai* (flies). The following words suggest that Philoponus means to canvass the notion that planet earth is a body in orbit, not stationary at the centre of the universe. It is ironic that Philoponus has chosen as his supposedly counterfactual – and patently absurd – premiss something that we now take for granted, along with the consequences that he also took to be crazy.

148. 'Subject to change', here and in what follows, translates the same term as 'in motion' in the classification of the first principles of earlier thinkers throughout the earlier part of this discussion.

149. On the motionlessness of souls see also below, 57,9.

150. The order of words in the Greek differs from the English. Philoponus says (literally) 'But then nor to resolve …' is in place of 'for nor to resolve …'. He is interested in how it supports the previous move in the argument, and whereas 'but then nor …' implies a new point, he holds that Aristotle is defending his procedure, for which the connective *gar* (for) would be normal.

151. Hippocrates of Chios, *c.* 470-400 BC. The most highly regarded geometer before Euclid, he is reported to have composed a book of 'Elements' anticipating Euclid in certain respects. His attempt at squaring the circle is more fully reported by Simplicius in his commentary on this passage of *Physics* Book 1, Simplicius *in Phys.* 60-8. For the details and bibliography see Bulmer-Thomas, 'Hippocrates of Chios'.

152. The task is to discover what square would have the same area as a given figure bounded by curved lines, in this case a circle, or (in effect) to discover any rectilinear figure, whose area can be calculated, that is equivalent in area to the curved figure and from which an equivalent square figure can thence be derived. Hippocrates succeeds in finding a solution for squaring some segments of the circle (lunes), but cannot infer, as he hopes to do, that we thereby have the answer for the circle.

153. Antiphon the Sophist, fifth century BC. The fragments are collected in Pendrick, *Antiphon the Sophist: The Fragments*. This passage is F13(d) in Pendrick's collection. For commentary on the differences between Philoponus' report here and those of Simplicius and Themistius see Pendrick, *Antiphon*, 262-7; 268.

154. There is inconsistency between the commentators as to whether Antiphon chose a square as the rectilinear figure to inscribe in the circle. See Pendrick, *Antiphon*, 262-7, who suggests that Philoponus derived his information from Alexander and did not use Eudemus' *History of Geometry*.

155. The corners of these multi-sided polygons have increasingly wide angles, which we might think of as greater angles rather than very small or slight ones (*mikras panu*), but the reasoning here is clear and the wording need not be a mistake if we take it to refer to the non-technical *look* of the figure: these are not sharp bends but very slight ones at the corners of a polygon that is close to the edges of the circle. See Rudio, *Der Bericht des Simplicius über die Quadraturen des Antiphon und des Hippokrates*, 107 and n. 2.

156. Philoponus does not necessarily imply that the words attributed here to Antiphon are a quotation. Rather they seem to be words 'put into his mouth' as an explanation of the construction with which he is credited.

157. Or more generally, 'curve'.

158. Philoponus' lemma here reads *autous* (it happens that they raise …) where most manuscripts of Aristotle read *autois* (it happens to them to raise …). The sense is not affected. Codex I has Philoponus' reading. Philoponus repeats the phrase (with *autous* again) in his commentary below at 32,8 and 32,10.

159. The Eleatics are said to say nothing '*peri phuseôs*', that is they contribute nothing in the field known as physics, or writing on nature. This is because they are not writing within that field, but criticising its principles from without. See above, 27-8.

160. This is a paraphrase and expansion of the next phrase in Aristotle's text, 185a20, serving as an informal lemma for exegesis.

161. i.e. metaphysics.

162. Following Vitelli's suggestion in the apparatus, I am reading *epeidê* in the lacuna before *ekhei* in line 23 and deleting it from line 25.

163. *pragmati* (in thing).

164. Philoponus refers to the person in the picture as 'the depicted person'. The context suggests that the issue is whether the person and the drawing of some person are one thing, given that both are referred to as a 'person'. The question is not whether the original person, and the character *portrayed* in a picture are the same individual, but whether the person and the picture are one. In the case of two individuals, each of whom is a person, they would be *really* one in respect of the universal. But the person and the drawing are not instances of the same universal, since one is a person and one is a drawing; they merely share the name 'person', and are thus *nominally* one but not *really* one.

165. Or 'because they have one name, the name of the person'.

166. Philoponus is building on Aristotle's discussion of the various senses of 'one' in *Phys.* 5.4, 227b3, where he introduces the notions of one in genus or one in species with similar examples. 'One in genus' and 'one in species' are translating *genei hen* and *eidei hen* respectively. The use of the entire class of substances as a typical example of being 'one in genus' is perhaps surprising.

167. The expression here is identical to that describing the case of the 'nominally one' above, 33,15-16.

168. *tên haplôs leukansin.*

169. See *Phys.* 185b7-9.

170. *lôpion* and *himation*; this example is given by Aristotle, *Phys.* 185b20.

171. Literally, two words in Greek (*aor* and *xiphos*) meaning 'sword'.

172. Substance *and* quantity is covered at *Phys.* 185b3 (discussed by Philoponus 39,30) and then substance alone at 185b4 (see 40,2-8 below).

173. *Phys.* 185a27-9. See 36,22.

174. 185a32.

175. *Post. An.* 1.22, 84a12-14: what belong *per se* are in two sorts: things that belong to them in the definition, and things to which they belong in the definition of those things. Aristotle's examples are 'odd' which belongs to number *per se*, because number is in the definition of odd, and 'multitude' or 'divisible' which belong to number *per se* because they belong in the definition of number. Philoponus expands Aristotle's rather cryptic dichotomy reasonably accurately, but he does not make clear where he changes from the first sort to the second (where I have inserted (b)) and he does not explain which sort he thinks applies in this case; it must in fact be an example of type (a), though he runs it straight on (at 34,32) from his explanation of type (b).

176. *hupokeimenon* (substrate), here the subject of predication.

177. Hinnible (capable of neighing). On this technical scholastic term and its human equivalent (risible, capable of laughing) see below, note on 57,5.

178. i.e. the first line mentioned, the one enclosing the figure.

179. Infinity belongs to quantity *per se* in accordance with the first kind of *per se* belonging (that is (a) above) because quantity is a component in the definition of infinity.

180. *Cat.* 6, 5b1-2.

181. *Phys.* 185b5.

182. See above 33,24-6, and (b 2.1), (b 2.2), and (b 2.3) on the diagram.

183. Philoponus' reading *poteron hôs ousian hapanta ê posa tauta ê poia* (whether as substance all these things or quantities or qualities ...) diverges in a number of details from the main manuscript tradition for Aristotle, and it seems likely that *tauta* (these) is an error since the repetitions of the lemma below (36,6.12) have *atta* (certain) at that point. Ross's text reads *poteron ousian ta panta ê posa ê poia* (whether all substance or quantities or qualities ...).

184. For a discussion of earlier contributions to interpreting this sentence, see Simplicius *in Phys.* 71,19-75,9. Simplicius disapproves of Alexander for looking to find the universal section of the classification in Aristotle's treatment, and favours Porphyry's reading which takes this sentence to raise the distinction between what is nominally one and what is really one. Perhaps Alexander is among Philoponus' 'some people' here.

185. The 'universal section of the classification' is the section dealing with what is one in respect of the universal, see above 33,17, and (b 1) on the diagram.

186. See above, 33,17-19, and (b 1.1) on the diagram.

187. *Phys.* 185a23-4. Philoponus has *de panta* (but all) where most Aristotle manuscripts (and Ross) have *ta panta* (all of them).

188. See above 33,19-23, and (b 1.2) on the diagram.

189. i.e. the classification of what is one in respect of the particular; see 33,23, and (b 2) on the diagram above.

190. *Phys.* 185b7-9.

191. See above, 33,23-6; 35,28-36,2; 36,8-9. Cf. 40,21; 45,16; 67,5-7.

192. See above, 34,2, and (a) on the diagram.

193. I have found it impossible to make sense of the reading *kath' hauto*, and have translated on the basis that the text should read *hôsper hai deka katêgoriai kat' auto to on hen eisin*.

194. For the account of the two arms of the classification see 33,12-26. Here Aristotle returns to the second alternative, (b) on the diagram, namely that things are 'really one' – which was the subject before the digression onto 'nominally one' (185a27), explained by Philoponus at 36,21-37,5. The division into substance and attribute did not appear in the account at 33,12-26 but was added above at 35,26-7.

195. See above 33,30.

196. *Phys.* 185a30.

197. This phrase is excised by Vitelli: it appears again a few lines further down. However, I think it is wrong to excise it, since what follows is an argument for the claim that 'impossible' is a stronger form of 'absurd'. Philoponus states the claim and then reflects upon the relationship between impossible and necessary, absurd and plausible, before repeating the claim as he moves to the conclusion that impossible and absurd are degrees on the same scale. A similar analysis for the same claim is given by Simplicius (*in Phys.* 75,10-18) but Philoponus' fuller treatment is presumably the original.

198. *Phys.* 185b3 (below, 39,30).

199. 185a27.

200. Philoponus comments on the fact that *kath' hupokeimenou* (in relation to a substrate) occurs where we might expect *en hupokeimenôi* (in a substrate). Cf. Aristotle *Cat.* 1a20-b10. What is said in relation to a substrate includes secondary substances or universals, and Philoponus seems to suggest that all the other categories can only properly be in a substrate. In fact Aristotle does allow that some other things, besides substance, are said in relation to a substrate (e.g. *Cat.* 3a21). Quantity would presumably be like that, though Aristotle does not say so.

201. The text is somewhat corrupt in the manuscripts at this point. I have translated, slightly loosely, the reading given by Vitelli which seems to be the best-supported, but it would seem more natural to read *autos* ('Melissus himself suggested it was quantity') in place of *auto* ('Melissus suggested that it was itself quantity').

202. The sentence would read better without *kath' hauto* ('*per se*'). If '*per se*' is retained, then *ei mê kata sumbebêkos* must be translated 'but only *per accidens*'.

203. See above, 35,3-8.

204. *ousiôdeis.*

205. 'Additional to the essence', *epousiôdeis.*

206. Philoponus was to have more to say on the idea that three-dimensionality might be the essential property of body in later treatises. This forward looking hint towards the need for a dedicated work on the subject is an important indicator that he is already (in perhaps 517) toying with the idea (which was to become prominent in his *De Aeternitate Mundi contra Proclum* of 529). Notice that here he does not endorse for himself the claim that body is composite and that matter is an integral part of body, but rather suggests that Aristotle is committed to that view (which thus explains why Aristotle does not take the route Philoponus has just described, regarding infinity and quantity). On the chronology see further below, 55,26 and note, and compare above, 4,28 and note.

207. *hêitini* is read as a single word.

208. *sunekhê* should be read without iota subscript.

209. Again Philoponus leaves room for the idea that three-dimensionality is perhaps constitutive of the essence of bodies; but three-dimensional extendedness is to be distinguished from the body's size, which is an accidental property of the body. It is size, not three-dimensionality, that is said to be finite or infinite.

210. See above, 33,8-16.

211. The word 'being' in this lemma occurs in all but one of the MSS of Philoponus' version and in the thirteenth-century Vaticanus 241 MS of Aristotle, but is absent from the remainder. The sense is not at issue since 'being' is automatically understood from the context.

212. See above, 33,4-9.

213. See above, 33,11-26; cf. 35,25-36,2.

214. *Phys.* 185a27. See above 36,22.

215. 185a29. See above, 37,5.

216. 185a32. See above, 34,15-35,16; 38,11-39,29.

217. 185b4.

218. cf. *Phys.* 5.3, 227a10-12.

219. *kinêma* is the term for a partless change or 'move': cf. Aristotle *Phys.* 241a4. See Simplicius *in Phys.* 1027,18-21 and Philoponus *in Phys.* 861,5-6.

220. See *Cat.* 10a11-16.

221. Philoponus gives two poetic words for a mortal person, *merops* and *brotos*. There is no precise English equivalent.

222. *Phys.* 185b20.

223. *logos.*

224. Neither Aristotle nor Philoponus explicitly quotes any relevant text from Heraclitus here (but cf. *Topics* 159b30, which suggests that Aristotle thinks that *Phys.* 185b21 is a close paraphrase of something in Heraclitus). The extant fragments that best fit this description are perhaps DK 22B58-62 and B126.

225. I am reading *autai men gar hai* ... (for the ideas themselves) rather than *hautai men gar hai* ... (for these things, the ideas).

226. This description of Zeno's motives and the structure of his argument corresponds with the account put into the mouth of Zeno in Plato's *Parmenides*, 128C-E.

227. The discussion here probably refers to the argument attributed to Zeno (DK 29A16) to the effect that if anyone could show him what the one was, Zeno would be able to speak of 'entities' (*onta*). The argument is cited in that form by Simplicius *in Phys.* 97,12-13 (commenting on this passage of Aristotle). Simplicius is quoting from Eudemus' discussion, also a commentary on this part of Aristotle's *Physics*. In recent literature, the argument is often taken alongside the arguments from physical division (since if one divides any proposed unit into further units one cannot say how many items there are), but Philoponus and Simplicius both take the comment to relate to the fact that an individual can be classified in many ways, according to the various attributes and parts into which one can subdivide any supposed individual entity – as in Philoponus' claims here about the fact that Socrates is both white, pot-bellied, snub-nosed, and a philosopher, for instance. So one cannot have a list of entities in the world unless one first delimits what counts as one thing. Philoponus takes the subdivision of a continuum into spatial parts to be a second, distinct, argument (summarised below, 43,1-4). He responds to the two arguments with separate solutions at 44,31 and 45,2.

228. 'He' in this sentence continues to refer to Zeno, even though the choice of example (Socrates), by means of which Philoponus cashes out the Zenonian reasoning, seems a trifle anachronistic.

229. This may be a summary of the argument explained (and apparently quoted verbatim from Zeno) by Simplicius at *in Phys.* 140,27-141,8 (DK 29B3 and B1).

230. Diels diagnosed a lacuna here. Vitelli proposes to fill the gap as follows: 'since it is impossible for there to be several units, of which the plurality is composed, hence ...'.

231. Lycophron the Sophist: possibly the butt of Plato's joke against late learners at *Sophist* 251B, to which Philoponus alludes below, 49,23. Testimonia are collected in DK 83.

232. See below, 49,19, where Philoponus identifies this allusion as being to Menedemus of Eretria (*c.* 339-265 BC).

233. i.e. *hupokeimenon*, what underlies.

234. Clothing and garment: *lôpion* and *himation*, see 33,27; ground and land: *khthôn* and *gê*.

235. The example above, 44,14-15.

236. Above, 42,18-28.

237. See above, 43,26-7.

238. Above, 43,1-6.

239. See above, 43,27.

240. Three words in Greek meaning sword (*aor*, *xiphos*, *makhaira*). For all three classes and the same examples, see above, 33,23-6, and the diagram ad loc.

241. I am convinced that the text given by Vitelli (which I have attempted to translate, for want of a better one) is incorrect here: the phrase *kai hai zôai autai* (and the lives themselves) is both out of place and unexpected, and indeed uncharacteristic vocabulary for the context. I suspect that the reading of K (*autou* in place of *autai*) is correct, and that *hai zôai* perhaps hides a feminine participle, ending *-zousai*, parallel with *sumplêrousai*, presumably *idiazousai*.

The resultant text (*hai Sôkratous idiotêtes hai sumplêrousai auton kai idiazousai autou, ei kai mê eisi kath' hautas sunekheis* ...) would be translated 'the characteristics of Socrates that go to make up him, and that are individuating of him, even if not in themselves continuous, nevertheless have their being in something continuous'.

242. Here Philoponus seems to fail to follow his own advice (just developed at length) to the effect that something might be one in various respects (and indeed one in some respects and several in others). He tries to find a single definitive answer to how Socrates counts as one, whereas he should have allowed that he might be one in various ways (and several in other ways).

243. The Lydian is one of the standard modes of ancient Greek music. The choice of this mode, and the details of the chord under discussion, are not significant to the example.

244. This probably refers to 45,25-46,21, since the issue is not addressed or mentioned in the Exposition and Discussion section relating to these lemmata (i.e. Section 3.2.1 above).

245. The sense demands the translation I have given. It is obtained by reading (as Vitelli suggests in the apparatus) *apêrtêmenê* (disconnected) in place of *anêrtêmenê* (connected).

246. The translation of the lemma is deliberately literal to reproduce the difficulties of interpretation that Philoponus is discussing. Philoponus takes it that the second phrase gives a consequence of assuming the first condition true, rather than the justification for why we might suppose the first condition likely.

247. See the diagram in Section 3.1.1.

248. Above, 41,1-6.

249. *lôpion* and *himation*: see above 33,27.

250. i.e. (b 2.3). For the three suggestions see *Phys.* 185b7 and above 33,25-7; 45,18, and the diagram in Section 3.1.1.

251. The reference to Heraclitus is Aristotle's (185b20). Philoponus explains Aristotle's cryptic allusion by interpreting it as a reference to Heraclitus' unity of opposites thesis. See above, 41,15, for a fuller account of the point.

252. Clearly intended to be a quotation from Aristotle's text, but Philoponus' reading does not precisely correspond with any of the manuscripts of Aristotle, which seem to say 'and to be good and not good'.

253. *Phys.* 185b22-3.

254. Menedemus of Eretria, *c.* 339-265 BC, who established a Socratic school at Eretria. See Diogenes Laertius 2.125-44. Since Aristotle died in 322, the dates seem wrong for Aristotle to be referring to someone who was only 17 when Aristotle died, let alone to identify him among 'the more recent of the ancients'. It is possible that Philoponus (and/or any previous authorities to whom he may be indebted here) confused Aristotle's reference to Lycophron with the second century Lycophron who was an associate of Menedemus of Eretria. See Simplicius *in Phys.* 91,28; 93,32 for vague references to 'folk from Eretria' – references which may go back to Eudemus.

255. See Simplicius' discussion at *in Phys.* 99,25-31 which confirms that Philoponus is probably talking about Alexander of Aphrodisias. Simplicius diagnoses Alexander's suggestion (to the effect that Plato was the one who 'remodelled the language') as deriving from a misunderstanding of a point originally made by Eudemus.

256. In these statements *esti* (is, there is) stands emphatically at the beginning of the statement. They could be translated 'the just <u>is</u> something', etc.

257. Plato *Sophist* 251B.

258. *Phys.* 184b25-186a3.

259. The idea seems to be to reconstruct an account of Melissus' views that is as sensitive and charitable as possible and in accordance with his original intentions, before looking at whether Aristotle's refutation is fair. This does not, of course, guarantee that Philoponus' attempt at sensitive presentation is successful.

260. The 'some' in this comment may include Eudemus: see Simplicius' discussion of the reasoning at *in Phys.* 104-5.

261. cf. Melissus fragment 1 (DK 30B1), quoted by Simplicius *in Phys.* 162,23-6.

262. The method in syllogistic logic that converts a premiss into its converse; e.g. all that is grass is green, nothing that is grass is not green. The two versions of the premiss are equivalent, but we cannot obtain 'all that is not grass is not green' by conversion of 'all that is grass is green'. Whether Melissus correctly converts the premiss is a matter discussed below (52,12 ff.).

263. i.e. conversion by substitution of the contradictory (e.g. not-green for green).

264. In 'all that is grass is green', 'all that is grass' is the antecedent and 'is green' is the consequent. We can negate the consequent 'what is not green' and add the negation of the antecedent 'is not grass' without a problem, but if we negate the antecedent 'what is not grass' and then negate the consequent 'is not green' we get a false result unless the two terms were co-extensive (i.e. if all and only grass were green).

265. Philoponus illustrates his point. 'Human' is the antecedent, 'animal' the consequent. The two terms are not coextensive (because there are animals that are not human). Hence if we negate the antecedent (if not human) and draw a conclusion about the consequent (then not animal) the conclusion comes out false. It will be correct, however, to negate the consequent (if not animal) and draw a conclusion about the antecedent (then not human).

266. The text looks to be corrupt, lacunose, or jumbled. This paragraph begins by indicating that there is a valid syllogism in the second figure, obtained correctly by conversion from the consequent. This syllogism is apparently then illustrated in the sentence 'If what comes to be has a beginning then what does not have a beginning did not come to be' (i.e. if A then B; so if not B then not A, by conversion from the consequent). We then need to supply a further premiss (missing or implied) 'Being does not have a beginning' (not B), followed by the conclusion at 52,25, 'Being therefore did not come to be' (therefore not A). This, however, is a sound syllogistic sequence. In the text as it stands we now have one further repetition of the converted premiss, 'What does not have a beginning did not come to be' (if not B then not A), 52,25-6, followed by the unexpected claim that the sequence does not make a syllogism. Two explanations seem likely: (1) The text originally illustrated both the correct method of conversion from the consequent, generating the argument given at 52,24-5 and also Melissus' incorrect method of conversion from the antecedent (of which most is now lost and only the premiss at 52,25-6 and the judgement that it is asyllogistic remain). Or (2) the repeated premiss at 25-6 is an error by dittography and the judgement that the sequence is asyllogistic is an observation made by a subsequent reader faced with the incoherent text.

Reconstructing Melissus' invalid argument (see further below, 53,2-3), we must suppose that we start with the premiss 'What comes to be has a beginning' (if A then B). 'What comes to be' is the antecedent and 'has a beginning' is the consequent. Melissus converts by negating the antecedent (if it does not come to be, if not A) and drawing a conclusion about the consequent (it does not have a beginning, then not B), in order to generate support for the claim that because being did not come to be, therefore it has no beginning (if not A then not B) whereas the conversion is reliable only if we negate the consequent (if it does not have a beginning, if not B) and draw a conclusion about the antecedent (it did not come to be, then not A).

267. The term *arkhê* in Greek has two senses, one of which is conveyed by our word 'beginning', used to translate *arkhê* in the preceding discussion of Melissus' argument. The other relates to our notion of a first, or governing, principle, which may (as for the early natural philosophers) be a material element, referred to later as the 'elemental principle', 59,31, or may be the governing part as in the examples of parts of the body given here, and referred to later as the 'partile principle' (59,28). This also may (but need not) occur at the beginning in time, as in the case of the blastocyst. 'Real' (*pragmatikê*) here in the sense of something that is related to a thing or object. This distinction is used again below, 56,24 and 59,15-24.

268. This term is the modern scientific equivalent of what in Philoponus' primitive embryology is called *sarkion* (i.e. speck of flesh).

269. Philoponus, unlike Aristotle, does think that the universe has an origin, so he, unlike Aristotle, has an interest in showing that Melissus was also mistaken in his arguments for the impossibility of being having a beginning. On Philoponus' developing views on this subject see Sorabji, 'John Philoponus', 23, and Verrycken, 'Development'.

270. The term *mus* in Greek had, like our 'mouse' only more so, a number of other meanings besides the small rodent, e.g. a mussel, a kind of whale, muscle of the body.

271. I have attempted to translate *tosauta ... hêgeisthô* (MS) as 'such [arguments] are to be advanced'. Vitelli suggests *arkeitô* (let them suffice) and Diels (followed by Verrycken, who, however, translates as if he read *arkeitô*) suggests *eirêsthô* (let them be mentioned). The notion of 'sufficing' is the least satisfactory, since the next phrase suggests that the sufficient arguments are provided elsewhere, and hence that these are in some sense not adequate, but all that we ought to offer at this point. I prefer to keep the manuscript reading, despite its oddity.

272. This back reference to an earlier work on the origin of the world is a problem for the traditional dating of the *Physics* commentary to 517, before the main works on the eternity of the world. It is unclear what earlier work of his own Philoponus means to indicate here. Verrycken argues in favour of *De Aeternitate Mundi contra Proclum* (Verrycken, 'Development', 252-4), which is dated 529. Other candidates are *De Aeternitate Mundi contra Aristotelem* (for which see Wildberg, *Philoponus, Against Aristotle on the Eternity of the World*) and the lost third treatise known as *De Contingentia Mundi*, both normally dated after 529. For Verrycken's revised dating, suggesting a post-529 second redaction of the *in Phys.*, see Verrycken, 'Development', 244-54. It now seems unlikely that the reference is to the lost *Summikta Theoremata*, as was suggested by Evrard, 'Convictions religieuses'.

273. *pros alêtheian*: the preposition *pros* (towards) may mean no more than 'with reference to' or 'relating to', but Parmenides conceives of the enquiry into truth as a journey to a place so that the directional sense seems appropriate. For this reason I have treated these words and the later 'towards opinion' (*pros doxan*) as the titles of the parts of the poem, often known as the 'Way of Truth' and the 'Way of Opinion'.

274. Earlier (22,2) Philoponus had given fire and earth as the two principles in the *pros doxan*. This corresponds with what Aristotle claims elsewhere (*Metaph.* 986b34), and may be what Philoponus should have written here.

Neither coincides with the received text of Parmenides, which has fire and night.

275. See above, 22,10.

276. It is not immediately apparent who 'the man himself' is, given the plural 'these men' at the start of the paragraph and the intervening parenthesis about Parmenides. Philoponus does not explicitly return to discussion of Melissus' reasoning until 57,3, and the intervening Neoplatonic interpretation of the realms of being and becoming is more convincing as an explanation of the background intentions of Parmenides (as the context in which Melissus was working).

277. The plural here presumably picks up the plural 'these men' at 55,27, despite the singular 'of the man himself' in the previous sentence.

278. The identity of 'him' may be Melissus (the long term target of this passage) or Parmenides (who has just been associated with Melissus as part of the same Monist project).

279. Plato *Timaeus* 27D6-7.

280. Plato *Timaeus* 28A1-3.

281. For the distinction between chronological *arkhê* and real *arkhê* see above 52,27-53,2 and below 59,15-24. Since the term *arkhê* is ambiguous in this way, it is not possible to give a single indecisive translation for it. It is translated by 'beginning' for the temporal *arkhê* and 'principle' for the real *arkhê*. In some cases it is unclear which sense is intended.

282. On the complaint that Melissus converted invalidly from the antecedent, see above, 52,12 ff. and notes ad loc.

283. The technical term 'risible' here is the scholastic term of art traditionally used to translate *gelastikon* (capable of laughing). It is the correlate, in relation to humans, of 'hinnible' (*khremetistikon*, capable of neighing) in relation to horses: all and only humans are capable of laughing and all and only horses are capable of neighing, so humans are risible animals and horses are hinnible animals. This sense of 'risible' has a longer pedigree in English than the more familiar sense (namely, something that excites laughter). For 'hinnible' used in this way, see above, 34,27.

284. The reference is presumably to Aristotle *Phys.* 8.6, 258b10-260a19; cf. also *Phys* 2.7, 198a17 (*akinêta* used of mathematicals). See also above, 30,16-18 on the changelessness of souls.

285. The Greek word meaning 'nothing' literally means 'not one', hence the reasoning is that if what is not is not one, then what is must be one. The suggestion here is that this involves an incorrect conversion from the antecedent: whereas Parmenides infers from 'What is apart from what is is not one' to 'what is is one', he should correctly have inferred from 'What is apart from what is is not one' to 'What is one is <in> being'.

286. In fact the falsity of the premisses may also be irrelevant to the truth of the conclusion, since a true conclusion may yet be derived from a combination of false premisses and an invalid argument.

287. See *SE* 165b26-166a21, and *Top.* 148a23-b23.

288. i.e. the part of a thing that is the governing part. See above 52,27 ff.

289. The phrase translated above, in the lemma, 'Just as the part that is inside ...'. Philoponus is explaining that we get this reading by reading '*enon*' (being inside) rather than '*hen on*' (being one). The difference depends upon substituting a rough breathing and grave accent on the letter *e* in place of a smooth breathing and no accent. When Aristotle wrote, no marks were shown for accents, rough or smooth breathings, or divisions of words, so the text could be read either way.

290. The phrase is now read with *hen on* (being one) taken as separate words (and with a rough breathing imagined on *en* to make *hen* (i.e. one)).

291. A reminder that Aristotle's suggestion that there is nothing to prevent it changing is taken to be merely dialectical. See above 53,30-54,7, where Aristotle is said to share the view that the infinite is unchanging, but to refute Melissus on the grounds that his proof (from infinity alone) is inadequate. Philoponus holds that Melissus is right to make an inference from strict unity to immutability. He does not explicitly criticise Aristotle, but he seems to suggest that Aristotle fails to see that Melissus bases his argument on this correct inference, and hence that Aristotle's refutation is an *ignoratio elenchi.*

292. Theophrastus *Phys. dox.* fr. 7 Diels (= FHSG number 234). Theophrastus is also cited for this argument by Simplicius *in Phys.* 115,11-15; 118,2-3; 134,11-12.

293. Perhaps a reference to *Sophist* 237D-E.

294. See above, 58,24-33, where Parmenides was also said to have two arguments. The first (argument a) is formulated by Philoponus at 58,6-8. This is the argument also found in Theophrastus. The second (argument b) is formulated by Philoponus at 58,32-3. Neither argument is explicitly cited by Aristotle.

295. i.e. the first argument, argument a.

296. See above 58,7; 58,24.

297. The 'conclusion' is the preliminary conclusion implicitly derived from the first two premisses and then converted to give the conclusion actually expressed. The errors are more fully explained above, 58,7-13. On conversion from the antecedent, see above 52,12 and notes ad loc.

298. The material in this paragraph largely repeats material covered in the previous section (Section 5).

299. See *Sophist* 241D. Much of Plato's *Sophist* is devoted to clarifying the ways in which something can be said not to be, and to refuting Parmenides' arguments against the reality of not-being.

300. See above 33,16-23.

301. For the threefold options see above, 36,7-9; *Phys.* 185b7-9.

302. *Phys.* 185b20. See above, 41,12-25 and 49,3-4.

303. *sumbebêkos.*

304. Vitelli diagnoses a lacuna at this point. It may be possible to read the sentence as undamaged if slightly chaotic.

305. *Phys.* 186a34-b4.

306. 186b4-14.

307. Or '... will not be being'.

308. *Phys.* 185a22.

309. Parmenides, DK 28B8, line 4.

310. This line appears to be a variant or misquotation of what we know as DK 28B8, line 5 (excerpted from Simplicius *in Phys.* 78,14 and 145,5).

311. The second half of DK 28B8, line 25.

312. This brief survey of a few lines of Parmenides' poem swiftly demonstrates to Philoponus' satisfaction that Parmenides granted that there was more than one being in the world. He therefore concludes that the unity that Parmenides asserted was the unity of the intelligible world, in accordance with Philoponus' Neoplatonic reading of the poem, which takes the *Towards Truth* to be focused on the intelligible world. Philoponus does not develop his analysis, but he is implicitly rejecting the discussion in Aristotle which treats the poem as asserting that there is just one being.

313. i.e. the arguments against Melissus at *Phys.* 186a4-22.

314. See above, 50,22-61,22.

315. Philoponus' comment here (Text 707 in Gigon, *Aristotelis Librorum Deperditorum Fragmenta*) is the only evidence for Aristotle's work *Against Parmenides*. Cf. Zeller, *Die Philosophie der Griechen*, vol. 1, p. 621, n. 2. We do not know to whom 'people' in 'people say' might refer.

316. *Phys.* 185a9-10. Cf. 186a7-10.

317. See above 62,7-63,7 for Philoponus' earlier analysis of the premisses of the argument and his objections.

318. On the error of conversion, see above 52,12 and notes ad loc. Cf. also 58,7-13.

319. *Phys.* 186a27.

320. 186a28.

321. See above, 35,28-36,2; 36,8-9; 40,21-4. The reference is to *Phys.* 185b7-9.

322. i.e. as the sole meaning of 'being', in place of 'white' which has been serving as Aristotle's illustration (for the sake of investigating the effect of taking being to be just an attribute).

323. Above, 63,20-2.

324. These words (being-as-such) are missing from the manuscript, and are supplied by Vitelli.

325. *Phys.* 186a34-5.

326. 186a35.

327. 186b2-3.

328. The alternative ways of construing this sentence are discussed below: for this reason I have given a non-committal translation showing both alternatives. Modern editors of Aristotle sometimes emend the text to resolve the issue.

329. Punctuating after *mallon* (rather) and reading *ti* as 'something' (i.e. 'something else rather') with the dative *ekeinôi* (of that) and understanding the verb *sumbebêken* again.

330. The question begins not with *ti* (why) but with *to hoper on* (being-as-such).

331. To obtain this reading we punctuate after *ekeinôi* (of that) and understand it to mean 'of nothing except of that thing itself'. Then *ti* (why) is read as the beginning of the question and *mallon* (rather) as part of the question ('why rather ...?').

332. *Phys.* 186b9-10.

333. Philoponus' reading of the text (reading 'if' after 'hence') is found in the main family of Aristotle's manuscripts. A variant without 'if' is found in Simplicius, the versio Arabo-Latina, and the first hand of MS E, and is preferred by Ross.

334. This translation is designed to retain the ambiguity discussed by Philoponus below at 75,20-6.

335. Possibly a reference to *Republic* 509B, where the Form of the Good is said to be above and beyond being; but cf. also *Timaeus* 28C.

336. See *Timaeus* 51A-B.

337. Again apparently a reference to *Republic* 507A-509C, where the sun is given as an analogy for the Good, which is itself indescribable directly.

338. See *Timaeus* 50A-B, where, however, the material suggested is gold.

339. *onomatôn, ê onomatôn kai rhêmatôn*. Philoponus uses the general word *onomata* (names) to refer to linguistic terms generally, and then uses it in the more precise sense (nouns or noun phrases) as distinct from predicate terms or verbal expressions.

340. cf. above, 33,26.

341. Separable attributes are those that can be absent from a subject without altering the identity or definition of the subject. Inseparable attributes are essential to what it is to be that thing.

342. The point is that not only can the attribute be defined without reference to the subject, but also the subject can be envisaged without the attribute in question.

343. The subject (*hupokeimenon*) here seems to refer not to an individual entity that has the attribute (e.g. the number three) but a kind of thing (number). The definition of odd or even must incorporate a reference to number, though not to any particular number.

344. On *per se* properties see above, 34,22-35,3 and notes, and below, 74,16-18. They come in two kinds: attributes which include their subject in their definition, and attributes which belong in the definition of their subject. Snubness and even are *per se* properties of the first kind, since nose is part of the definition of snub and number is part of the definition of even.

345. Philoponus finds this axiom in *Phys.* 186b34; see 78,15-25 below.

346. Or 'of being'.

347. Or 'being'.

348. Or 'being itself'.

349. Apparently the first axiom listed above, 72,31-73,1.

350. This problem seems to be envisaged for human being and its attributes but not for number, presumably because 'even' is not a part of number (in which case to include number in the definition of even would be to introduce the whole in a definition of the part) but every bearer of the attribute 'even' is a number, and hence number is invariably the subject and is included in the definition, even though, just as not all animals are human beings, so not all numbers are even. The difference seems to be that human is treated as the subject of animal, and animal as a part or attribute of human, whereas number is treated as the subject of even and even as an attribute of some numbers. Were animal the subject of which human was a predicate then animal, entering the definition of human, would be precisely parallel to number entering the definition of even, since all evens are numbers and all humans are animals. Philoponus treats 'animal' as a part of a human being, and hence suggests that we cannot use human being in our definition of animal for that reason. His difficulties probably arise from the attempt to treat attributes that are not inseparable in the first sense but only in the second sense, namely the items attributed to human being in the definition, as *ex hypothesi* inseparable.

351. That is, inseparable ones of the first sort, in which there is only one kind of thing to which they can belong.

352. See the list of axioms, 72,30-73,16.

353. See above, 34,22-35,3.

354. See above, lines 18-19.

355. See 73,14-16.

356. 'either': this is the second alternative (attributes of something else) refuted (this alternative was canvassed at 74,24). The first alternative (attributes of the entity itself) was refuted in two limbs (separable or inseparable attributes) at 73,17-74,24. For the two main options, see 72,28-9.

357. *Phys.* 186b12.

358. Sentences of subject-predicate form can be formulated with the logical subject in the grammatical predicate position, and *vice versa*. Philoponus says such propositions are *para phusin* (contrary to nature) because they do not capture the logical structure of the reality (in which attributes belong to subjects, not subjects to attributes).

359. 76,15-16.

360. The point here seems to be that neither the logical elements of the definition nor the physical elements of the physical body include the whole in their own definition.

361. The last two remarks are slightly puzzling here, partly because the grammatical structure changes so that 'bed' and 'Socrates' are in the nominative case and need to be the subject of a verb which is not supplied (I have supplied 'is incorporated'). Also it is far from clear that we can define 'foot of the bed' without mentioning 'bed' in the definition, and if it is Socrates' hand that is to be defined, as opposed to hand in general, it might seem that we would mention Socrates in defining it. Presumably 'foot of the bed' is here serving as a part from which the whole bed is composed, viewing the whole bed as a composite of parts, rather than of the four elements; the question is whether it is right to say that the whole never enters the definition of a specialised part of this sort. Philoponus does not appear to see a problem.

362. See above 73,14-16 and below 78,15-25.

363. The words 'and each' are put into square brackets by Vitelli, who appears to favour excising them on the grounds that Philoponus did not acknowledge their presence in his earlier discussion (see 73,14-16 and 74,28-9). They occur in some but not all manuscripts of Aristotle's text. But in what follows Philoponus discusses whether they should be included, so that it is evident that they were noted as a variant reading in his text, along with other variants (see below 78,20-5). Simplicius likewise knows that text as a variant reading (*in Phys.* 129,25).

364. This is the point at which Philoponus suggests that Aristotle is employing the fourth of the axioms listed above; see 73,14-16 and 74,28-9.

365. On this variant reading see the note to the lemma.

366. Philoponus implies that this variant reading (*kai katholou*) is an alternative to 'and each' (*kai hekateron*), but it appears that its place in the sentence would actually be to replace 'and that of which' (*kai kath' hou*), though it may still be the case that it was not found combined with 'and each'. The resulting sentence on this variant reading would be 'But let a being as such be what is not an attribute of anything, and in general let it be said to be both and the product of these'.

367. Ross and others read this sentence as a question, but Philoponus clearly assumes that it is the conclusion of the argument.

368. See above 36,2; 40,22 ff.

369. Or 'being'.

370. Although Philoponus confusingly calls this the second argument, it is not the second but the first of the two covered by Aristotle's 'both arguments' in this lemma. Philoponus classifies it as the second of Parmenides' arguments because (in his view) it is a new one and not a repeat of the one discussed above (61,22-79,24, on 186a22-b35). Modern commentators (Ross, Charlton) assume that the reference is to the argument sketched and criticised above, the argument from the univocity of 'being'. Philoponus identifies a new argument with two premisses, i.e. besides the univocity of being, a further premiss to the effect that contradictions cannot simultaneously be true.

371. I am assuming *athakatos* is a misprint for *athanatos*.

372. Philoponus is here describing the argument commonly known as the Dichotomy or Stadium, which is also mentioned by Aristotle at *Phys.* 6.9, 239b11; 239b19 and rejected by him at *Phys.* 8.8, 263a4-264a6 (and see also [Aristotle] *Lin. Insec.* 968a18). All the ancient commentators agree that Aristotle is here referring to Zeno's argument 'from dichotomy'. Simplicius *in Phys.* 138,3 traces this explanation back to Alexander. Modern commentators agree that Aristotle alludes to Zeno, but Ross implausibly identifies a different Zenonian argument, not the one normally known as the argument from dichotomy (cutting in two). Charlton, however, agrees with Philoponus.

373. For Alexander of Aphrodisias see the quotation in Simplicius' discussion on this topic (Simplicius *in Phys.* 134,19-32). Simplicius also cites Porphyry (Simplicius *in Phys.* 135,1-14). Simplicius himself denies that Plato is intended, 135,15 ff., and supports his claim using material that he found in Alexander (which, we must presume, was not regarded by Alexander as an adequate defence). Ross (followed by Charlton) suggests (with reference to *GC* 324b35-325a32) that the atomists are more plausibly intended, in respect of their acceptance of the existence of absolute not-being (in the form of void) alongside a single kind of being, their reasoning being based on a naïve acceptance of the 'uniqueness of being' argument of Parmenides. For Themistius, see Themistius *in Phys.* 12,12-24.

374. Plato *Sophist* 258B. Cf. Simplicius *in Phys.* 136,10 ff. who paraphrases *Sophist* 258B and quotes 258E7.

375. This paraphrase is more a summary of the general enterprise in the *Sophist* than a quotation from any part of it.

376. I have translated *hoion ei tukhoi zôion ê ousia* though I find the phrase rather surprising, given that *ousia* does not seem to be a classic or helpful example of a genus. The reading of K (*hê ousia*) would give us 'as if substance were animal perhaps'. The difference lies only in an accent and breathing on the letter *ê*, and either reading could be preferred.

377. See above 33,13; 63,8.

378. Read thus as an objection to Plato's *Sophist*, Aristotle's points seem unfair, since Plato's response to Parmenides is precisely to identify a way of not being that equates with 'being other than', so that *being other than* a horse would be a way of *not being* that does not amount to *not being tout court*. Although Philoponus makes this point at 82,25, he also appears to endorse the idea that Plato meant to agree that what is really real, the intelligible being, is in fact numerically one (81,30-82,2) so that Plato both accepts the Eleatic conclusion that true being is one alone, and also rejects it by saying there are other things that are not simply non-existent. It seems likely that Philoponus has in mind some Neoplatonic interpretation which retains the idea that true being is just one (the One), but also makes room for a derivative kind of being for the plural intelligibles. Philoponus accepts without demur the testimony of Alexander and Themistius that Plato's *Sophist* is in Aristotle's sights here. Simplicius rejects the idea.

379. The argument from dichotomy, above 80,23-81,16.

380. See Themistius *in Phys.* 12,6-9; Simplicius *in Phys.* 138,3-18 (quoting Alexander of Aphrodisias).

381. Again Philoponus accepts the judgement of his predecessors that Xenocrates is intended, but Ross suggests that here too the atomists' theory might be under attack, this time for capitulating on indivisible magnitudes.

382. Philoponus sees some clever pattern in the idea of pupils and teachers capitulating to the arguments of pupils and teachers, but the point carries no convincing weight in support of his identification of the characters intended by *enioi* (some) at 187a1.

383. This recapitulates the point made at 81,29-82,5.

384. Philoponus notes that the word '*de*' (but) is at the start of a new point in the argument, and hence need not have a strongly adversative sense. His point is correct: see Denniston, *The Greek Particles*, 162. On his interpretation, which he here claims is his own suggestion, it is not necessary to read this lemma as an explanation of the previous point (for which the word '*gar*' (for) would be required – hence the previous paragraph which gives an interpretation that does require this sense), but it should instead be read as a wholly new point, which adds a further attack on Plato for his method of proceeding against Parmenides, which (we now learn) was not only mistaken but also unnecessary, according to this reading.

Bibliography

Barnes, J., *Early Greek Philosophy* (Harmondsworth: Penguin, 1987)

Bulmer-Thomas, Ivor, 'Hippocrates of Chios', in Charles Coulston Gillispie and Frederic L. Holmes, eds, *Dictionary of Scientific Biography* (New York: Scribner, 1972), vol. 6, 410-18

Charlton, William, *Aristotle's Physics Books I and II*, Clarendon Aristotle Series (Oxford: Clarendon Press, 1970)

Denniston, J.D., *The Greek Particles*, 2nd edn (Oxford: Clarendon Press, 1954)

Diels, H., and W. Kranz, *Die Fragmente der Vorsokratiker*, 3 vols (Berlin: Weidmann, 1951)

Evrard, Étienne, 'Les convictions religieuses de Jean Philopon et la date de son Commentaire aux Météorologiques', *Bulletin de l'Academie royale de Belgique, Classe des lettres, sciences morales et politiques, série 5* 39 (1953), 299-357

Evrard, Étienne, *L'école d'Olympiodore et la composition du 'commentaire à la physique' de Jean Philopone* (diss., Liège, 1957)

Fortenbaugh, W.W., P.M. Huby, R.W. Sharples, and D. Gutas, *Theophrastus of Eresus: Sources for his Life, Writings, Thought, and Influence*, 2 vols (Leiden: Brill, 1992)

Gigon, O., *Aristotelis Opera Volumen Tertium: Librorum Deperditorum Fragmenta* (Berlin: De Gruyter, 1987)

Guthrie, W.K.C., *A History of Greek Philosophy*, 6 vols (Cambridge: Cambridge University Press, 1962-81)

Haas, F. de, 'Philoponus on Theophrastus on Composition in Nature', in J. van Ophuijsen and M. van Raalte, eds, *Theophrastus: Reappraising the Sources* (New Brunswick: Transaction, 1998), 171-88

Inwood, Brad, *The Poem of Empedocles*, 2nd edn (Toronto: University of Toronto Press, 2001)

Kirk, G.S., J.E. Raven and M. Schofield, *The Presocratic Philosophers*, 2nd edn (Cambridge: Cambridge University Press, 1983)

Laks, André, 'Le début d'une physique: ordre, extension et nature des fragments 142-144A/B de Théophraste', in J. van Ophuijsen and M. van Raalte, eds, *Theophrastus: Reappraising the Sources* (New Brunswick: Transaction, 1998), 143-67

Long, A.A., 'The Principles of Parmenides' Cosmogony', *Phronesis* 8 (1963), 90-107

Long, A.A., and David N. Sedley, *The Hellenistic Philosophers*, 2 vols (Cambridge: Cambridge University Press, 1987)

Mourelatos, Alexander P.D., *The Route of Parmenides* (New Haven: Yale University Press, 1970)

O'Brien, Denis, *Pour interpreter Empédocle* (Leiden: Brill, 1981)

Osborne, Catherine, *Rethinking Early Greek Philosophy* (London: Duckworth, 1987)

Osborne, Catherine, 'Heraclitus', in C.C.W. Taylor, ed., *From the Beginning to Plato*, Routledge History of Philosophy, vol. I (London: Routledge, 1997), 88-127

Pendrick, Gerard J., *Antiphon the Sophist: The Fragments*, Cambridge Classical Texts and Commentaries (Cambridge: Cambridge University Press, 2002)

Ross, W.D., *Aristotle's Physics: A Revised Text with Introduction and Commentary* (Oxford: Clarendon Press, 1936)

Rudio, F., *Der Bericht des Simplicius über die Quadraturen des Antiphon und des Hippokrates*, Urkunden zur Geschichte der Mathematik in Altertume (Leipzig: Teubner, 1907)

Sorabji, R., 'John Philoponus', in R. Sorabji, ed., *Philoponus and the Rejection of Aristotelian Science* (London: Duckworth, 1987), 1-40

Sorabji, R., 'The Ancient Commentators on Aristotle', in R. Sorabji, ed., *Aristotle Transformed* (London: Duckworth, 1990), 1-30

Verrycken, Konrad, 'The Development of Philoponus' Thought and its Chronology', in R. Sorabji, ed., *Aristotle Transformed* (London: Duckworth, 1990), 233-74

Wildberg, C., *Philoponus: Against Aristotle on the Eternity of the World*, Ancient Commentators on Aristotle (London: Duckworth, 1987)

Wright, M.R., *Empedocles: The Extant Fragments* (New Haven: Yale University Press, 1981)

Zeller, Eduard, *Die Philosophie der Griechen*, 6th edn, vol. 1 (Leipzig: Reisland, 1919)

English-Greek Glossary

absolute: *haplos*
absolutely: *haplôs*
absurd: *atopon*
accidental, be: *sumbebêkenai*
actually; in actuality: *energeiâi*
addition, be in: *sunuphistasthai*
adjuncts: *parakolouthounta*
aloft: *meteôros*
alteration: *alloiôsis*
ambiguity: *homônumia*
analogy: *analogia*
angle: *gônia*
antecedent: *hêgoumenon*
appropriate: *eulogos*
arbitrary: *matên*
argue for; argue in support: *kataskeuazein*
argument: *logos*
arithmetic: *arithmêtikê*
arm/limb (of classification): *morion*
articulated: *diêrthrômenos*
assembly: *sunthesis*
assertion: *kataphasis, logos apophantikos*
assume (hypothetically): *hupotithenai*
assumption: *axiôma, hupothesis*
assumption, common: *koinon axiôma*
assumption, hypothetical: *hupothesis*
atmosphere, in the: *meteôron*
atmospherical phenomena: *meteôron*
atom: *hê atomos*
atomic: *atomos*
attributes: *sumbebêkota*
auxiliary cause: *sunaitios*
axiom: *axiôma*

become: *gignesthai*
before us: *prokeimenos*
beginning: *arkhê*
being: *to on*
being-as-such: *hoper on*
belong: *huparkhein*
birth: *genesis*
black; blackness: *melas; melania*
blackening: *melansis*
blastocyst: *sarkion*
bodiless: *asômatos*
body: *sôma*
boundary: *horos, peras*

cause: *aitia, aition*
censure: *memphesthai*
challenging: *antilogia*
change, subject to: *kinoumenos*
change: *genesis, kinêma, kinêsis, metabolê*
characterised, be: *kharaktêrizesthai*
characteristic (adj.): *kharaktêristikos*
characteristic: *idiotês*
characteristic, most: *idikôtatos*
chord: *harmonia*
classification: *diairesis*
clear: *dêlos*
clear-cut: *asunkhutôs*
clearer: *saphesteros*
coextensive, be: *exisazein*
coming-to-be: *genesis*
commentators: *hupomnêmatistai*
common: *koinos*
common assumption: *koinon axiôma*
common notions: *koinai ennoiai*
completion: *apotelesma*
component: *sustatikos*
concept: *logos*
conclusion: *sumperasma*
concurrence: *sundromê*
consequence: *akolouthoun, hepomenon*
consequent (in conditional): *hepomenon*
continuous: *sunekhes*
continuum: *to sunekhes*
contradiction: *antiphasis*

contradictory (n.): *antiphasis*
contrary: *antikeimenon*
contrary, be: *antikeisthai*
conversion by negation: *hê sun antithesei antistrophê*
conversion: *antistrophê*
correlative: *pros ti*
corruption: *phthora*
craftsman (feminine = nature): *dêmiourgos*
create: *dêmiourgein, poiein*
creation: *dêmiourgia, genesis*
creative: *dêmiourgikos*
creator: *dêmiourgos*
criticise: *elenkhai, elenkhein, enkalein*
curved: *kampulotês*

decay: *phthora*
deduce: *sunagein*
definition: *horismos, horos*
definitional specification: *horistikos logos*
definitive: *horistikos*
demolish: *anaskeuazein*
demonstrate: *apodeiknunai, deiknunai*
demonstration: *apodeixis, deixis*
demonstration, carry out a: *deiknunai*
demonstrative: *apodeiktikos*
denial: *apophasis*
deny: *anairein*
description: *hupographê*
detached: *apolelumenos*
detached, be: *apoginesthai*
development: *genesis*
dialectic: *dialektikê*
didactic: *didaskalikos*
didactic method of demonstration: *didaskalia*
differentia: *diaphora*
difficulty: *aporia*
diminution: *meiôsis*
discipline: *methodos*
discrete: *diôrismenon*
discrete (parts): *aphestêkota*
distinctive character: *idiotês*
diversification: *diakrisis*
division: *diairesis, dikhotomia, tomê*
doctrine: *dogma*

earlier thinkers: *arkhaioteroi, hoi palaioteroi*
education: *didaskalia*
efficient: *poiêtikos*
element: *stoikheion*
eliminate: *anairein*
enclose: *periekhein*
encompass: *periekhein*
entity, some: *ti on*
envisage: *epinoein, huponoein*
error: *planê*
essence: *ousia*
essential: *ousiôdês*
evidential: *tekmêriôdês*
evidentially: *tekmêriôdôs*
examine: *exetazein, prokheirizesthai*
exchange: *antapodosis*
exclusive: *idios*
exist: *einai, huparkhein*
existence: *huparxis*
expert: *epistêmôn*
explain: *epiluein*
explore: *gumnazein*
expression: *phônê*
extended in three dimensions: *trikhêi diastaton*

faculty: *dunamis*
fallacious: *asullogistos*
fallaciously: *asullogistôs*
false: *pseudos*
familiar: *gnôrimos*
fight against: *diamakhesthai*
figure (in syllogistic or geometry): *skhêma*
final: *telikos*
finer particles, made of: *leptomeresteros*
finite: *peperasmenos*
follow: *sumperainein, sumperainesthai*
force: *dunamis*
form: *eidos, idea, morphê*
format (v.): *eidopoiein*
form-giving: *eidopoios*
function: *energeia, logos*

gaseous, become: *exaerousthai*
general: *katholou, koinos*
general, more: *koinoteros*
generation: *genesis*
genus: *genos*
geometry: *geômetria, geometrikê*
growth: *auxêsis*

handiwork: *dêmiourgia*

hinnible (i.e. able to neigh or whinny): *khremetistikos*
homoiomeries: *homoiomerê, homoiomereia*
human; human being; human person; humankind: *anthrôpos*
hypothesis: *hupothesis*
hypothesis, be the: *hupokeisthai*

idea: *idea*
identify: *ginôskein, gnôrizein, sêmainein*
immanent: *sunkatatetagmenos*
immediate: *amesos*
immediately: *amesôs*
immutability: *to ametablêton*
impasse: *aporia*
impression: *phantasia*
inarticulate: *adiarthrôtos*
inarticulate manner, in an: *adiarthrôtôs*
incomplete: *atelês*
indefinite way, in an: *adioristôs*
indeterminate: *aoristos*
indeterminate manner, in an: *aoristôs*
indicative (v.): *horistikos*
indiscriminate: *sunkekhumenos*
individual: *atomon, kath' ekaston*
individual (adj.): *idios*
indivisible: *adiairetos*
infer: *sumperainein, sunagein*
infinite: *apeiros*
infinitude: *apeiria*
inhere: *sunuphistasthai*
innate heat: *emphuton thermon*
instrumental: *organikos*
intellective: *noeros*
intelligence: *noêsis*
intelligible: *noêtos*
intermediate (n.): *to metaxu*
invalid: *asullogistos, asumperantos*
invalidly: *asullogistôs*

juxtaposition: *parathesis*

kind: *genos*
know: *eidenai*
know, come to: *ginôskein*
know, get to: *gnônai*
knowable, more: *gnôrimôteros*
knowledge: *epistêmê, gnôsis*

lecture (v.): *didaskein*
lesson: *didaskalia*
letter: *stoikheion*
limitless: *apeiros*
line: *grammê*
linguistic: *di' onomatos*
literate: *grammatikos*
lucid: *enargês*
lune: *mêniskos*

magnitude: *megethos*
mathematical: *mathematikos*
mathematical science: *mathêmatikê epistêmê*
matter: *hulê*
mean; have a meaning: *sêmainein*
meaning: *sêmainomenon*
mental image: *phantasia*
metaphysician: *prôtos philosophos*
method: *methodos*
monad: *monas*
more knowable: *gnôrimôteros*
most characteristic: *idikôtatos*
motion: *kinêsis*
motion, in: *kinoumenos*
motionless: *akinêtos*
moulded, easily: *euplastos*
musical: *mousikos*

name: *onoma*
natural philosopher: *phusikos*
nature: *phusis*
necessity: *anankê*
necessity, every; entirely necessary: *anankê pasa*
nominally: *onomati*
non-essential: *epousiôdês*
not unreasonably: *eikotôs*
note: *êkhos*
notion: *ennoia*
noun: *onoma*
number: *plêthos*

objects: *pragmata*
obvious: *enargês, prodêlos*
obvious, what is: *enargeia*
obviously: *enargôs*
occur: *gignesthai, huparkhein*
one (numerically): *arithmôi hen, arithmôi mia*
one as such: *hoper hen*
one in form; one in species; one in respect of species; specifically one: *eidei hen, hen kata to eidos*

one in genus; generically one; one in kind: *genei en*
opinables: *doxasta*
opinion: *doxa*
opinion, established: *endoxon*
opposite: *enantios*
opposites: *enantia*
origin: *arkhê, genesis*
otherness: *heterotês*
own (adj.): *idios*

paradigmatic: *paradeigmatikos*
part: *morion*
participate: *metekhein*
particular (adj.): *idios, merikos, to kata meros*
partless: *amerês*
peculiar: *idikos, idios, kharaktêristikos*
peculiar characteristics; peculiarities: *ta idia*
penetrate through and through: *khôrein hola di holôn*
per accidens: *kata sumbebêkos*
per se: *kath' auto*
per se property: *pathos kath' auto*
perceptible: *aisthêtos*
perception: *aisthêsis*
person: *anthrôpos*
phoneme: *phônê*
physical theory: *phusiologia*
physical things: *phusika pragmata*
physicist: *phusikos, phusiologos*
physics, do: *phusiologein*
place: *topos*
plane: *epipedon*
plank: *xulon*
plausible: *eulogos*
plural: *polla*
plurality: *plêthos, polla*
point: *sêmeion*
posit (v.): *hupotithenai*
positing: *hupothesis*
posterior: *husteros*
postpone: *hupertithesthai*
potential; potentiality: *dunamis*
predecessors: *hoi palaioteroi*
predicate: *katêgoroumenon*
premiss: *protasis*
present (adj.): *prokeimenos*
present (v.): *proballein*
principle: *arkhê*
principles of physics: *phusikai arkhai*

prior: *proteron*
pristine: *katharos*
privation: *sterêsis*
problem: *aporia*
process of becoming: *genesis*
project: *prokeimenon*
project, be (someone's): *prokeisthai*
proof: *apodeixis*
proper: *idios*
property: *pathos*
proposition: *apophansis, prokeimenon, protasis*
prove: *apodeiknunai, deiknunai, elenkhai, elenkhein, kataskeuazein*
provide proof: *apodeiknunai*
providence: *pronoia*
proximate: *prosekhês*
punctuate: *hupostixai, stixai*
purely: *eilikrinôs*
puzzle: *aporia*

qualifications, without the: *haplôs*
quality: *poion, poiotês*
quantity: *poson*
question: *problêma*

ratio: *logos*
rational: *logikos*
real: *alêthinos, huparkhos, pragmatikos*
real being: *ontôs on*
reality, in: *en huparxei, en hupostasei*
really: *pragmati*
really real: *ontôs on*
reason: *aitia*
reasonably enough; not unreasonably: *eikotôs*
reasoning: *sullogismos*
recognise: *eidenai, ginôskein, suneidenai*
refute: *anairein, dialusai, dielenkhein, elenkhai, elenkhein*
relative: *pros ti*
reproductive, what is: *gonimon*
resist: *enstênai*
resolve: *epiluesthai, luein*
rest: *êremia*
risible (i.e. able to laugh): *gelastikos*
room: *khôra, topos*
rough manner, in a: *holoskherôs*

science: *epistêmê*

seeds: *spermata*
segment: *tmêma*
self-evident, what is: *enargeia*
self-evident: *autopistos, prodêlos*
self-supporting: *authupostatos*
seminal fluid: *gonê*
sense (thing signified): *sêmainomenon*
sentence: *logos*
separate: *khôristos*
several: *polla*
several ways, be said in: *pollakhôs legetai*
shape: *morphê, skhêma*
shoddy: *phortikos*
show (v.): *deiknunai*
sight: *opsis*
significance: *sêmasia*
signify: *sêmainein*
simple: *haplos*
simpliciter: *haplôs*
size: *ektasis, megethos*
snubness: *simotês*
solution, be the: *epiluesthai*
something that encompasses: *periektikon*
soul: *psukhê*
sounding: *apêkhêsis*
specialist: *idios*
species: *eidos*
specification: *logos*
specification, definitional: *horistikos logos*
spontaneous: *ek t' automatou*
struggle: *diateinomenoi*
study: *didaskalia, theôria*
subject: *hupokeimenon, pragmateia, theôria*
subject matter: *hupokeimenon*
subsist; have subsistence: *huparkhein, huphestêke, huphistêmi*
substance: *ousia*
substrate: *hupokeimenon*
suggest: *hupotithenai*
suggestion, be the: *hupokeisthai*
suggestion: *hupothesis*
suitability: *epitêdeiotês*
supervene: *epigignesthai*
support (v.): *kataskeuazein*
suppose: *huponoein*
supposed, be: *hupokeisthai*
surface: *epiphaneia*
syllogism: *sullogismos*
synonymously: *sunônumôs*

tackle (v.): *enstênai*
task, be (someone's) present: *prokeisthai*
teach: *didaskein, paradidômi*
teacher: *didaskalos*
teaching: *didaskalia*
temperament: *krasis*
term: *horos, onoma, phônê*
test (v.): *elenkhai, elenkhein*
theory: *dogma, logos, theôrêma*
tout court: *haplôs*
truth: *alêthes*
try: *gumnazesthai*

uncuttable: *atomos*
underlying: *hupokeimenos*
understand: *epistasthai*
understanding: *epistêmê*
unhypothetical: *anupothetos*
unification: *henôsis*
uniform parts: *homoiomerê, homoiomereia*
unintelligible: *adianoêtos*
unity: *to hen*
universal: *katholou*
universe: *ouranos, to pan*
unmoved: *akinêtos*
usage: *lexis*
utterance: *phônê*

verb: *rhêma*
void (n.): *to kenon*

white; whiteness: *leukos, leukotês*
whitened, be: *leleukôsthai*
whitening: *leukansis*
word: *onoma*
world: *kosmos*

Greek-English Index

Subject Index

Index of Passages

www.ingramcontent.com/pod-product-compliance
Ingram Content Group UK Ltd.
Pitfield, Milton Keynes, MK11 3LW, UK
UKHW020146290926
15471UKWH00004B/719

9 781472 557698